SHADED RELIEF MAP OF
ISLE ROYALE NATIONAL PARK, MICHIGAN
(PREPARED BY A. J. BURGESS)

The Geologic Story of
ISLE NATIONAL

By N. King Huber

Formerly U.S. Geological Survey Bulletin 1309

ROYALE
PARK

Revised by Isle Royale Natural History Association in cooperation with the National Park Service and the Geological Survey.

1983

UNITED STATES DEPARTMENT OF THE INTERIOR

GEOLOGICAL SURVEY

Library of Congress Cataloging in Publication Data

Huber, Norman King   1926-
The geologic story of Isle Royale National Park.

(Geological Survey Bulletin 1309)
Bibliography: p. 64-66
Supt. of Docs. No.: I 19.3:1309
1. Geology–Michigan–Isle Royale National Park–Guide-books. I. Title. II. Series: United States. Geological Survey. Bulletin 1309.
QE75.B9 [QE126.I8]   557.3'08s [557.74'997]   75-619126

Printed in the USA by Avery Color Studios, Marquette, Michigan 49855

ISBN 0-932212-89-1   First Edition - April 1983
                             Reprinted - 1986, 1990, 1996

For Isle Royale Natural History Association
Houghton, Michigan 49931

CONTENTS

The island in history	2
A look at the landscape	5
The building blocks—rocks and minerals	8
Volcanic rocks	8
Sedimentary rocks	13
Pyroclastic rocks	14
The way the rocks are stacked up	14
A volcanic pile—the Portage Lake Volcanics	14
Rock Harbor–Tobin Harbor area	17
Washington Harbor area	20
An ancient alluvial fan—the Copper Harbor Conglomerate	21
Tilting and breaking the rocks	29
Isle Royale—a small piece of the big puzzle	34
What happened when?	35
Time as a geologic concept	36
What was there before—the pre-Keweenawan	37
The rocks of Isle Royale—the Keweenawan	38
The missing chapters	40
The glaciers take over	41
Glacial effects on the land	41
Glacial effects on lake levels	47
Some minerals of special interest	55
Native copper—widespread but not abundant	56
Chlorastrolite—Michigan's state gem	58
Prehnite—the little pink pebbles	59
Agate—an array of colors	60
What the future holds?	61
A word of thanks	61
Definition of terms	62
References and additional reading	64

> To describe a biota there is no substitute for a sample. But it is logical to ask, what might one want to know which would require the preservation of a sample? Whether such a question is asked at all is a reflection of the stage of intellectual maturity of a civilization. We take it for granted that there is some social gain in the erection and maintenance of a museum of fine arts, a museum of natural history, or even a historical museum. Sooner or later we ought to be mature enough to extend this concept to another kind of museum, one which you might call the museum of land types, consisting of samples as uninfluenced as possible by man.
>
> *Luna Leopold, "The Meaning of Wilderness to Science," 1960: San Francisco, Sierra Club, p. 33.*

The Geologic Story of Isle Royale National Park

N. King Huber

Isle Royale is an outstanding example of relatively undisturbed northwoods lake wilderness. But more than simple preservation of such an environment is involved in its inclusion in our National Park System. Its isolation from the mainland provides an almost untouched laboratory for research in the natural sciences, especially those studies whose very nature depends upon such isolation.

One excellent example of such research is the intensive study of the predator-prey relationship of the timber wolf and moose, long sponsored by the National Park Service and Purdue University. In probably no other place in North America are the necessary ecological conditions for such a study so admirably fulfilled as on Isle Royale. The development of a natural laboratory with such conditions is ultimately dependent upon geologic processes and events that although not unique in themselves, produced in their interplay a unique result, the island archipelago as we know it today, with its hills and valleys, swamps and bogs—the ecological framework of the plant and animal world.

Even the most casual visitor can hardly fail to be struck by the fiordlike nature of many of the bays, the chains of fringing islands, the ridge-and-valley topography, and the linear nature of all these features. The distinctive topography of the archipelago is, of course, only the latest manifestation of geologic processes in operation since time immemorial. Fragments of geologic history going back over a billion years can be read from the rocks of the island, and with additional data from other parts of the Lake Superior region, we can fill in some of the story of Isle Royale. After more than a hundred years of study by man, the story is still incomplete. But then, geologic stories are seldom complete, and what we do know allows a deeper appreciation of one of our most naturally preserved parks and whets our curiosity about the missing fragments.

THE ISLAND IN HISTORY

Geology influenced the role that Isle Royale played in both pre-historic and historic activities of man in the Lake Superior region. Native copper implements, weapons, and ornaments have been found in Indian mounds and habitats throughout central and eastern North America, some of which probably date back to 3,000 B.C. Undoubtedly, most of this copper came from the Lake Superior region, and perhaps a significant part from Isle Royale. Even today, hundreds of Indian mining pits can be seen scattered throughout the park. The major concentration of these pits is near McCargoe Cove, where many were reexcavated later by miners and archaeologists. A charred log recovered from one of these pits yielded a radiocarbon age of about 1,500 B.C., attesting to the antiquity of these early mining operations.

The existence of native copper in North America was known to the European explorers as early as Cartier's second voyage in 1535–36. It was said to have come from the direction of the Lake Superior region. Not until the Jesuit missionaries penetrated the region more than a hundred years later, however, did more definite information as to the source of the copper become available, but even much of that was inaccurate. In the "Jesuit Relation" for 1669–70, a Father Dablon cites Indian legends regarding "an island called Menong [Minong], celebrated for its copper," including a rather extravagant description of the abundance of the metal. Minong was the Indian general term for island but was specifically applied to Isle Royale; it has been perpetuated on the island as Minong Ridge. As exploration continued, additional information about the sources of the copper was accumulated, and actual prospecting on Isle Royale began in 1843, the year in which the Chippewas relinquished their claims to the island. By 1850 copper mining by Europeans was established on Keweenaw Peninsula and had been attempted on Isle Royale (fig. 1).

In 1847 the U.S. Congress authorized a survey of the vast Lake Superior Land District to determine its mineral potential, and the result of this survey was the first series of comprehensive reports on the geology of the region. Charles T. Jackson's report of 1849 contains the first map of Isle Royale to show in a very general way the different rock types present on the island. Two following reports, both by John W. Foster and Josiah D. Whitney in 1850 and 1851, include an improved geologic map of Isle Royale and a wealth of information on the early mining activities on the island. Whitney later became the first State Geologist of California, and the highest mountain in the conterminous United States is named for him. These three reports show an amazingly sophisticated understanding of the general geology of the region, especially considering the limited amount of time that was spent exploring such a vast and nearly trackless wilderness. Foster and Whitney lucidly describe some of the difficulties in working on Isle Royale:

> The physical obstructions to a successful exploration of the island are greater than we encountered in any other portion of the mineral district. The shores are lined with dense but dwarfed forests of cedar and spruce, with their branches interlocking and wreathed with long and drooping festoons of moss. While the tops of the trees flourish luxuriantly, the lower branches die off and

2

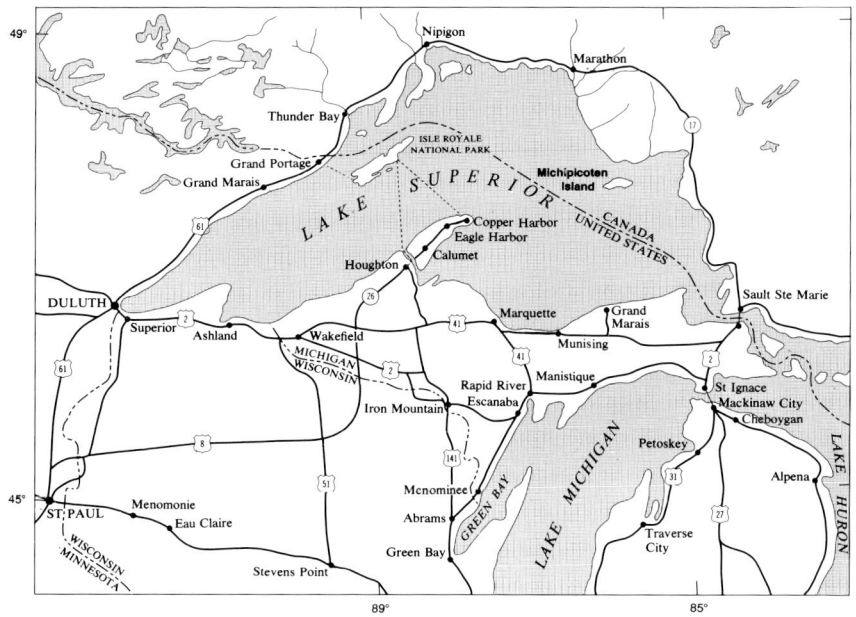

REGIONAL SETTING of Isle Royale National Park. (Fig. 1)

stand out as so many spikes, to oppose the progress of the explorer. So dense is the interwoven mass of foliage that the noonday sunlight hardly penetrates it. The air is stifled; and at every step the explorer starts up swarms of musquitoes, which, the very instant he pauses, assail him.

And so it remains today, but at least the modern explorer has relatively effective mosquito repellent available to him. Among the more astute conclusions reached by Jackson and by Foster and Whitney is that Isle Royale and the Keweenaw Peninsula may be regarded as geologic counterparts on the opposite sides of a structural basin occupied by Lake Superior. They also predicted correctly that "the island, for mining purposes, may be regarded as infinitely less valuable than Keweenaw Point."

The first period of mining activity on Isle Royale, from 1843 to 1855, ended with much disappointment, as no economic deposits were found. Mining was renewed during the period from 1873 to 1881 with the development of the Minong and Island mines. Although these two mines were the most productive ever to be opened on the island, they also were uneconomic, and this mining revival was short lived (fig. 2). Finally, from 1889 to 1893, an extensive exploration program was undertaken on the west end of the island, east of Washington Harbor. No economic deposits of copper were located, however, and the efforts were abandoned.

It was during this latest period of exploration that the most detailed and comprehensive study of Isle Royale to date was made by Alfred C. Lane of the Geological Survey of Michigan. His geologic map and report were published in 1898. The timing of this study was fortunate indeed, as samples ob-

HOIST AND BOILER REMAINS at Island mine, abandoned in 1877. (Fig. 2)

tained with diamond drills during the mineral exploration of this period were made available to Lane and permitted him to work out the succession of rock strata—the stratigraphy— of the island in much greater detail than would have been possible from the very incomplete sequence to be seen in natural rock exposures. The samples also permitted him to make stratigraphic correlations between Isle Royale and the Keweenaw Peninsula, which are for the most part still valid today. Lane's report was so comprehensive that until very recently only a few additional studies of limited geologic scope have been made on the island; most of these have been related to the glacial and post-glacial geologic history. Many other studies throughout the Lake Superior area, however, have given us a better understanding of the regional geologic setting within which we must view Isle Royale.

Recent field studies undertaken by myself, and Roger G. Wolff, have resulted in a series of technical publications on the geology of Isle Royale, which are listed at the end of this report. Included is a new geologic map (Huber, 1973c), published separately, which can profitably be used as a companion to this report. The map is at a sufficiently large scale (1 in.=1 mi.) to show details of the geology of Isle Royale not possible to show at the scale of maps in this report. In fact some of the geologic units described in this report are shown only on the larger map, which is also recommended for place names.

A LOOK AT THE LANDSCAPE

One's first impression of Isle Royale, when approaching by boat, is of an inhospitable rocky shoreline behind which a verdant mantle of vegetation softens what appears to be an irregular, hilly landscape. From this lake-level view, no topographic pattern is readily apparent (fig. 3). But if one approaches by air, a rather striking pattern becomes apparent, and one sees the island as a series of long parallel ridges and valleys, which give it a washboard aspect. This pattern is most conspicuous at the northeast end of the island where the ridges project into Lake Superior as so many rocky fingers and the drowned valleys between them have a distinct fiordlike appearance (fig. 4). The pattern is also dramatically displayed by the chains of elongate islands that fringe the main island—no more than the protruding tops of almost completely submerged ridges.

To the hiker traveling across the grain of the country, the series of ridges seems never ending; seldom walking level, he is monotonously going up and down. He also quickly becomes aware of another aspect of the topography—the southeast-facing slopes are relatively gentle compared with the northwest-facing slopes, which are often precipitous; the ridge-and-valley structure is decidedly unsymmetrical. And to his discomfort, the hiker finds that even when traveling with the grain of the island, he is still spending an inordinate amount of time going up and down, for the linear ridges are interrupted at frequent intervals by crosscutting ravines or depressions.

Another aspect of the topography that is readily apparent to the hiker is the generally poor drainage on the is-

ROCKY SHORELINE on Isle Royale. (Fig. 3)

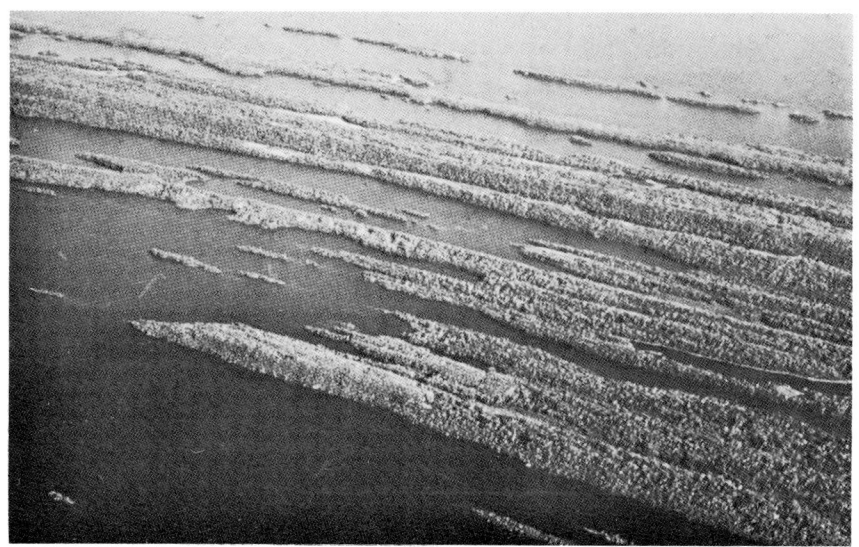

FINGERLIKE PROMONTORIES on northeast end of Isle Royale. Photograph by Alan Eliason, National Park Service. (Fig. 4)

land. Every valley seems to be at least partly occupied by swamps and bogs, as well as numerous lakes, all adding to the difficulty of traversing across the grain of the island. For those willing to expend the effort, however, hiking on Isle Royale can be an enormously rewarding experience, for the very topographic factors that make hiking arduous also provide numerous vantage points for some of the most scenic vistas in the Lake Superior region (fig. 5).

VISTA FROM LOOKOUT LOUISE looking across Duncan Bay. (Fig. 5)

And the sight of a bog mat tinged magenta with miniature orchids or a bull moose feeding in a beaver pond is not soon to be forgotten (fig. 6).

BULL MOOSE feeding in beaver pond. (Fig. 6)

The landforms of Isle Royale, like landforms everywhere, are the result of geologic processes and are dependent upon the type of underlying rocks, the structure of the rocks, and their erosional history. The bedrock sequence on Isle Royale consists of a thick pile of lava flows and sedimentary rocks that has been tilted toward the southeast, and the linear ridges of the island are the eroded edges of individual layers of the pile (fig. 7). Many geologic processes in combination worked together to create Isle Royale as we see it today, not only its landforms, but the rocks and minerals themselves. These processes and their consequences are discussed at some length on the pages that follow in an attempt to unravel the geologic story involved at Isle Royale National Park.

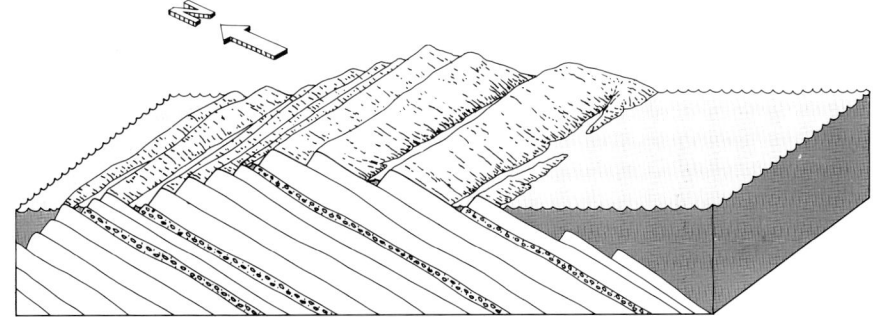

RIDGE-AND-VALLEY TOPOGRAPHY related to tilted rock structure. Circled areas indicate sedimentary rocks; blank areas indicate lava flows. (Fig. 7)

7

THE BUILDING BLOCKS — ROCKS AND MINERALS

VOLCANIC ROCKS

When we think of volcanic activity, we usually conjure up visions of a Vesuvius spewing forth billowing clouds of steam and ash or a Kilauea with fiery fountains of molten lava reaching skyward. This is partly because most people are familiar with the type of volcanism that builds imposing cones, such as Vesuvius, Kilauea, Mount Rainier, Mount Fuji, and many others and partly because many of these volcanoes have erupted in historic times, sometimes with disastrous results. Individual eruptions of such volcanoes, while spectacular, generally involve relatively small volumes of molten lava. In other words, such volcanoes are constructed slowly by frequent eruptions of relatively small quantities of lava spewed out in all directions from a generally localized vent. The lava seldom travels far from the vent, and the tongue-shaped flows accumulate one upon another to form a conical pile (fig. 8).

However, in some of the world's most extensive volcanic areas, true volcanoes are rare or absent. In such areas tremendous volumes of lava, sometimes measured in tens or even hundreds of cubic miles, welled out through fissures not connected with volcanic cones to form extensive sheets of flood basalts or plateau basalts. These sheets of basaltic lava, piled up flow upon flow to thicknesses of thousands of feet, are the most extensive of all volcanic deposits, and such deposits have occurred at various places and times in the geologic past. Such eruptions are rare at the present time, however; the only known historic eruptions of this type occurred in Iceland. There, at Laki, in 1783, an estimated volume of 3 cubic miles of lava flowed from a fissure 20 miles long to cover an area of 215 square miles. But even this was small compared with similar eruptions during earlier geologic periods.

The volcanic rocks of Isle Royale are flood basalts, but before discussing their details we can look at a younger sequence of flood basalts that is still relatively undisturbed geologically, allowing us to deduce more about their origin. In the northwestern United States, between 18 and 10 million years ago, enormous amounts of basalt lava welled up through deep fissures in the earth's crust and flooded much of present-day Washington, Oregon, and Idaho to form the great Columbia River Plateau. These basalt flows are known collectively as the Columbia River basalt (fig. 9).

Much of the Columbia River basalt has been eroded away, but it is estimated that the lava flood once covered an area of 150,000 square miles, equivalent to the combined areas of Washington and Oregon. It was probably over 1 mile thick in places and had

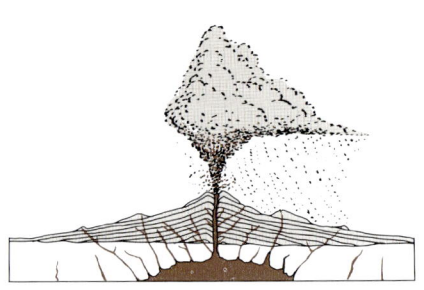

VOLCANIC CONE—cross section. (Fig. 8)

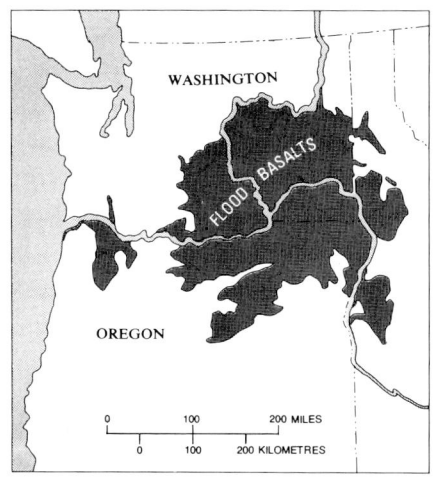

COLUMBIA RIVER BASALT—today. (Fig. 9)

FLOOD BASALTS of the Columbia River Plateau form these cliffs along the Yakima River. The Roza lava flow lies near the top of the hill. Photograph by R. S. Fiske. (Fig. 10)

an average depth of 1,800 feet. The tremendous amount of basalt contained in even a single flow can be illustrated by the Roza lava flow in Washington (fig. 10). This flow, which can be traced over an area of 8,800 square miles, averages 100 feet in thickness. Its volume of at least 160 cubic miles is roughly seven times the volume of Mount Rainier, one of the great volcanoes of the Cascade Range.

A schematic diagram of an eruption of flood basalt is shown in figure 11.

FLOOD BASALT ERUPTION—schematic diagram. (Fig. 11)

The lava wells out of long fissures that cut through earlier, cooled and solidified flows, and it is erupted in such a great volume that it rapidly flows away from the source fissures when it reaches the earth's surface. Instead of building volcanic cones, the eruption forms a vast lava lake, and the resulting flow, when solidified, tends to have nearly level surfaces. A sequence of such eruptions results in the accumulation of a series of basalt sheets covering vast areas—much like a layer cake—having quite a different geometry from the outward-dipping concentric structure that results from the eruptions of the more familiar cone volcanoes.

At first glance the volcanic rocks on Isle Royale seem to have a monotonous similarity—just nondescript dark-colored rocks. When studied more closely, however, numerous variations can be seen, some readily apparent and others more subtle. These variations are more significant than might be supposed. Some, for example, result in differences in rock hardness, which in turn affects resistance to erosion and thus the development of landforms. Other differences help to identify individual lava flows, thus enabling one to separate them from their neighbors and to trace them across the countryside. This recognition is critical to the preparation of a geologic map and to the unscrambling of the geologic history.

The compositional classification of volcanic rocks is largely based upon their silica (silicon dioxide) content, with a range from basalt, with a silica content of roughly 50 percent or less, to rhyolite, with a silica content of 70 percent or more. Rhyolite and volcanic rocks of intermediate composition usually have an excess of silica, and so some of the silica occurs in the free or uncombined state as quartz, as well as in other more complex silicate minerals. In basalt, however, all silica generally is needed to form silicate minerals, and quartz is uncommon.

The volcanic rocks on Isle Royale are nearly all basalt, and their mineral constituents are plagioclase feldspar, pyroxene, and lesser amounts of olivine, magnetite, and other minerals. A few of the rocks are somewhat more silicic than basalt and would be classified as andesite. Volcanic rocks as high in silica as rhyolite are not exposed on the island. In fact, the basalts of Isle Royale are so much alike chemically that we cannot readily use chemical composition by itself to distinguish one volcanic rock on the island from another.

When lava cools and solidifies quickly, it forms glass. If it cools more slowly, it partly or completely crystallizes to mineral grains. The glass, in time, will also crystallize to mineral grains, but they generally will be finer than those formed during slow cooling. Differences in the cooling history, as well as in chemical composition, of individual lava flows result in differences in rock texture, a property that chiefly reflects the grain size, shape, and distribution of the minerals in the rock. The basalts of Isle Royale, and elsewhere in the Lake Superior region, do exhibit variations in texture; consequently, a rock classification based upon textures has developed through the years and is widely used in the Michigan copper district. Only an experienced geologist can estimate the chemical compositions of rocks in the field, but anyone can learn to recognize the different textures; remembering their names is the major hurdle. The textural classification is extremely useful because it can be applied directly in the field, at the outcrop, and thus is a powerful aid in identifying individual flows for geologic mapping. I must emphasize, however, that these textural terms are not always used with the same connotation outside the Lake Superior region, and some geologists have abandoned them altogether because of conflicting usage. But they will be found in any further readings on the geology of the volcanic rocks of the Lake Superior region. The rock textures are illustrated in figure 12.

Ophite.—Rock with a mottled texture produced by crystals of pyroxene surrounded by a slightly darker matrix of finer grained minerals (fig. 12*A*). Pyroxene has good cleavage—the tendency to break along definite planes, producing smooth surfaces. Such surfaces reflect light better than the rock matrix, and on freshly broken specimens of ophite, the flashing of pyroxene cleavage surfaces in the sunlight is a distinctive feature. Cleavage

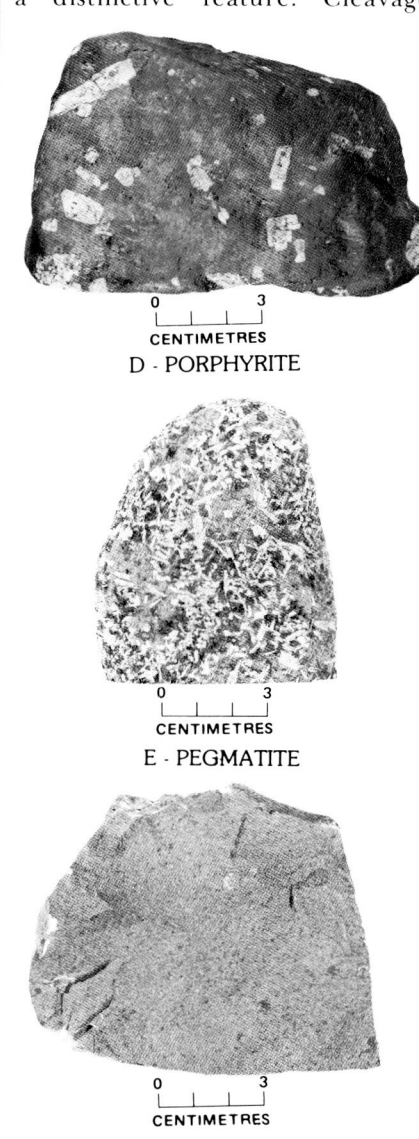

VOLCANIC ROCK TEXTURES found on Isle Royale. (Fig. 12)

reflections are dulled by chemical processes involved in weathering, but weathering actually accentuates the texture by increasing the color contrast between the pyroxene crystals and the rock matrix and by producing a knobby surface (fig. 13). The size of the pyroxene crystals varies from less than 1 millimetre to more than 2 centimetres, but crystal sizes are usually rather uniform in any individual specimen. Within a given lava flow, the pyroxene crystals are progressively larger toward the flow interior. Ophite is the most abundant volcanic rock type on Isle Royale.

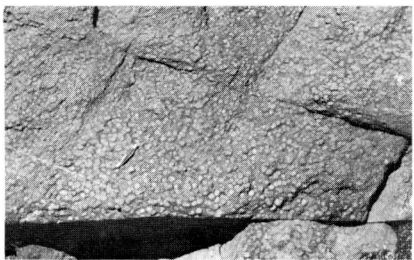

COARSE-GRAINED OPHITE showing weathered, knobby surface (Fig. 13)

Porphyrite.—Rock with a texture produced by well-defined plagioclase crystals scattered through a finer grained groundmass. The term is applied to two distinct varieties of such porphyritic rocks. One variety has small, blocky millimetre-sized plagioclase crystals rather uniformly distributed through the groundmass (fig. 12B); when the plagioclase crystals tend to clot together, the term glomeroporphyrite is used (fig. 12C). The other variety of porphyrite has larger, tabular-shaped crystals more sparsely distributed in the rock and commonly occurring in clots (fig. 12D); the large crystals are often as much as 2 centimetres long.

Pegmatite.—Rock with a texture in which all of the minerals, especially the plagioclase, are larger when compared with those in most of the other rock types; the elongate plagioclase laths give the rock a matted appearance (fig. 12E).

Trap.—Fine-grained dark-colored massive rock showing none of the distinctive textures mentioned (fig. 12F). Trap commonly breaks with a curved fracture.

Felsite.—Fine-grained light-colored volcanic rock, generally reddish. Texturally it is most similar to trap, although commonly containing scattered crystals, but is applied to more siliceous volcanic rocks such as rhyolite. Felsite does not occur in lava flows exposed on Isle Royale but does occur as pebbles in most of the conglomerate on the island.

All the rock textures are more easily recognized on weathered than on fresh surfaces because color contrasts between many minerals are increased during weathering and most lava flows can readily be characterized as ophite, porphyrite, or trap. In some flows the texture is obscure, but in nearly all these flows the rock is ophite. Pegmatite is a unique rock found only in layers in the interior of some ophitic flows, generally the thicker ones; its formation involves late-stage crystallization of a liquid remaining in the central part of a flow after the bulk of the flow has already solidified.

Two other terms are useful in describing the volcanic rocks: vesicle and amygdule. As a lava flow cools, small circular cavities are formed by the expansion of bubbles of gas or steam during the solidification of the rock; these cavities are called vesicles. When the cavities are later filled with minerals deposited from solutions circulating through the rock, the vesicle fillings are called amygdules. A rock with abundant amygdules is called an amygdaloidal rock or simply an amygdaloid (fig. 14); Amygdaloid Island, for ex-

AMYGDALOID with calcite amygdules and calcite vein at top. (Fig. 14)

ample, is named for a rock of this character. The rocks on Isle Royale are very old and have been saturated by mineral-bearing solutions long enough so that no vesicles remain—all have been filled and converted to amygdules. The uppermost part of most individual lava flows is conspicuously amygdaloidal, containing as much as 50 percent amygdules. The abundance of amygdules decreases downward toward the amygdule-poor massive basalt of the middle and lower part of the flow, with another usually much thinner amygdaloidal zone reappearing at the base of the flow (fig. 15). The

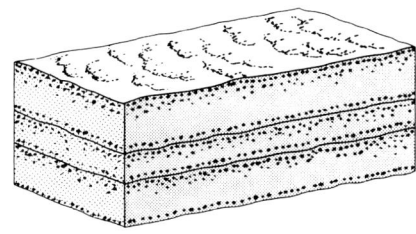

LAVA FLOWS—cross section showing distribution of amygdules. (Fig. 15)

amygdaloidal zones are less resistant and more easily eroded than the flow interiors and normally do not crop out as well. Because of this, the actual surface of a flow is only rarely observable (fig. 16).

ANCIENT LAVA FLOW with ropy surface (above) on Isle Royale compared with a "modern" lava flow (below) at Craters of the Moon National Monument. (Fig. 16)

SEDIMENTARY ROCKS

The sedimentary rocks of Isle Royale, sandstone and conglomerate, are easily recognized as being the lithified or consolidated equivalents of sand and gravel. These rocks exhibit a wide range in coarseness from a very fine grained sandstone to conglomerate with boulders 2 feet in diameter. Most of the erosional debris that formed the original deposits of sand and gravel was derived from volcanic rocks with a wide range in composition. A striking rusty-red color is characteristic of all the sedimentary rocks—a color that reflects the presence of a significant quantity of hematite (iron oxide) in the fine groundmass in which the larger fragments are imbedded. The fact that many of the volcanic rock fragments themselves have a reddish tone enhances this coloration.

PYROCLASTIC ROCKS

In contrast to the relatively quiet fissure eruptions of flood basalts, at various times more violent eruptions of rhyolitic lava hurled vast quantities of ash and other volcanic material into the air. The resulting rain of volcanic debris covered extensive areas and formed deposits that because of their fragmental character, superficially resemble sedimentary deposits. Rocks resulting from the consolidation of such volcanic deposits are known as pyroclastic rocks; those formed from very fine material, such as ash, are called tuffs. Pyroclastic rocks are not abundant on Isle Royale but do occur sandwiched between some of the flood basalt flows.

THE WAY THE ROCKS ARE STACKED UP

The bedrock sequence on Isle Royale consists of a thick pile of lava flows and sedimentary rocks that has been tilted toward the southeast. Having already looked briefly at the different varieties of rocks that occur in the pile, we can now examine the rock sequence in more detail and get to know the subtleties of its character and the relative position or succession of its component parts—its stratigraphy.

In geology, a formation is an assemblage of rocks that have some character in common, whether of origin or composition. It is thus a way of dividing rocks into manageable units for discussion or for depiction on a geologic map. The rock sequence on Isle Royale has been divided into two formations, one deposited above the other. The lower—and thus older—one has been named the Portage Lake Volcanics and includes all the lava flows and their minor interbedded sedimentary and pyroclastic rocks. The upper formation, the Copper Harbor Conglomerate, contains only sedimentary rocks; it is mostly conglomerate but also includes sandstone. On the Keweenaw Peninsula of Michigan, where these formations also occur and were originally named, other formations can be seen both above and below the Portage Lake Volcanics and Copper Harbor Conglomerate. Some of the other formations undoubtedly also extend to the vicinity of Isle Royale, but they are concealed beneath the waters of Lake Superior.

A VOLCANIC PILE— THE PORTAGE LAKE VOLCANICS

There are probably over 100 individual lava flows in the Portage Lake Volcanics on Isle Royale with a total thickness of over 10,000 feet—quite an impressive pile. And the total thickness of the formation must be greater because although the top of the formation is exposed, the bottom is not, and an unknown number of additional flows undoubtedly lie beneath Lake Superior on the north side of the island.

The sedimentary rocks interbedded with the lava flows are exposed only under unusual circumstances because they are more easily eroded than the volcanic rocks and as a result are generally buried beneath surficial materials in depressions between ridges of more resistant volcanic rock. We know from holes drilled during exploration for copper deposits that about 25 sedimentary layers 1 foot thick or more are present in the sequence; however, less than a third of these layers are known from outcrop, and all those are in the upper part of the sequence.

In order to construct a geologic map, which shows the geographic distribution of individual rock units and their relationships to each other, some method must be found of distinguishing key rock units from each other; usually, rock composition does the trick. But within a pile of similar lava flows, this is a distinct problem. Fortunately, some of the lava flows in the Portage Lake Volcanics can be distinguished from their neighbors above and below on the basis of their distinctive rock textures. These flows can then be used as divisions or marker units within the sequence of flows and can provide stratigraphic and structural control for depicting the sequence on a geologic map. The interbedded sedimentary rocks would also serve as marker units for mapping if they were exposed more frequently.

The amygdaloidal zones at the base and top of flows are less resistant than the massive flow interiors; therefore, the interior parts of the thicker flows are the dominant ridge-forming units. Contacts between flows, on the other hand, are seldom exposed as they are generally concealed beneath alluvial materials or talus in the depressions between ridges. Fortunately for geologic mapping, the distinguishing texture of an individual flow is generally better developed and more readily apparent in the nonamygdaloidal flow interiors, and therefore, that part of a flow most useful for purposes of identification is invariably the most likely part to be exposed.

The volcanic rocks on Isle Royale are predominantly ophitic, and except for a few unusual ophites, the nonophitic flows are most conspicuous among their neighbors and make the best horizon markers. The finer grained porphyrites and traps are also generally more resistant to erosion than ophitic flows of similar thickness and thus tend to stand out topographically, making it easier to trace them across the countryside and adding to their usefulness in geologic mapping. There are 12 flows within the Portage Lake Volcanics on Isle Royale that are distinctive enough to be easily separated from the rest of the flows in the pile, and these have been named for ease of reference (fig. 17). Most of these named flows are less than 100 feet thick, but six of them can be traced the length of the island. The named flows include five porphyrites, four traps, and three ophites. Only four of these individual flows are shown on the geologic map in this volume (pl. 2), but all are shown on the larger scale geologic map published separately (Huber, 1973c).

For most visitors to Isle Royale, the northeast and southwest ends of the island are the most accessible, although a growing number of visitors are hiking the length of the island on the Greenstone Ridge Trail, which follows Greenstone Ridge, the backbone of the island. At neither end of the island can all the named lava flows or volcanic rock types be readily seen, but enough variety is present to provide interesting geological excursions from Rock Harbor Lodge and Windigo Inn via small boats available at both places. A somewhat less varied assortment of rocks can be seen on short hikes from the lodge areas.

Named lava flows:		Unnamed rock sequences:
	Copper Harbor Conglomerate	
		Numerous thin ophitic flows with as many as seven interbedded sandstone and conglomerate units
Scoville Point Flow (porphyrite)		
		Several thin ophitic flows
Edwards Island Flow (trap)		
		Conglomerate—known from drill records
Middle Point Flow (porphyrite)		
		Several thin ophitic flows
Long Island Flow (trap)		
		Sandstone—known from drill records
Tobin Harbor Flow (porphyrite)		
Washington Island Flow (ophite)		
		Tuff-breccia
Greenstone Flow (ophite)		
		Conglomerate—known from drill records
Grace Island Flow (porphyrite)		
		Sequence of thin to thick (more than 100 ft) flows chiefly ophitic, with one or more sedimentary units suggested by drill data
		Tuff-breccia
Minong Flow (trap)		
		Sequence of thin to thick flows, chiefly ophitic, with one or more sedimentary units suggested by drill data
Huginnin Flow (porphyrite)		
		One or more ophitic flows present locally
Hill Point Flow (ophite)		
		Sequence of thin to moderately thick flows, chiefly ophitic. Several sedimentary units and a felsite indicated by drill records
		Breccia
Amygdaloid Island Flow (trap)		
		Lava flows, chiefly ophitic

PORTAGE LAKE VOLCANICS—schematic columnar section on Isle Royale. Not drawn to scale. (Fig. 17)

ROCK HARBOR— TOBIN HARBOR AREA

Most of the rock types in the Portage Lake Volcanics can be seen in the Rock Harbor—Tobin Harbor area (pl. 1). The main exception is the coarse porphyrite characteristic of the Huginnin Flow (fig. 12D), which can only be seen on the north side of the island.

All the volcanic rocks on the chain of islands on the south side of Rock Harbor are ophitic, and excellent exposures can be found on the wave-buffeted south shores of those islands. Ophitic texture is especially well developed on the south side of Raspberry Island where exposures are readily accessible via the Raspberry Island Nature Trail (fig. 13 is from this locality). Ophite can also be seen along the Commodore Kneutson Trail near the Rock Harbor Lodge.

Fine-grained porphyrite can be seen in the Scoville Point Flow at various points on the north shoreline of Rock Harbor, as at Threemile Campground, on Scoville Point, and on the south sides of Edwards and North Government Islands. Its speckled appearance is quite distinctive (fig. 12B). Rock with a somewhat similar texture (fig. 12C) can be seen in the Tobin Harbor Flow at various points along the north shoreline of Tobin Harbor, as near the dock for the Lookout Louise Trail, on several small islands in Tobin Harbor, and on the south side of Porter Island. The trail to Mount Franklin also surmounts a bold outcrop of the Tobin Harbor Flow shortly after crossing Tobin Creek.

Fine-grained trap of the Edwards Island Flow can best be seen on the small promontory just north of Scoville Point, on Split Island, and on the north sides of Edwards and North Government Islands. Trap of the Long Island Flow can be seen on several small islands opposite the dock for the Lookout Louise Trail and on Long and Third Islands. The Edwards Island Flow and to a lesser extent the Long Island Flow have an important secondary attribute that helps in their field recognition; except for the upper part of the Greenstone Flow, they are the only flows in the volcanic sequence that commonly exhibit well-developed columnar jointing (fig. 18).

COLUMNAR JOINTING IN EDWARDS ISLAND FLOW, Edwards Island, (Fig. 18)

Columnar joints are cracks that divide a lava flow into long vertical columns ideally tending toward a hexagonal cross section. They are formed during cooling of the flow under certain specific conditions. Rapid and uniform cooling rates tend to promote the development of columnar joints, but the degree of homogeneity of the solidifying rock is probably more important. It can be shown mathematically that the surface of a homogeneous medium should be divided by a crack system defining regular hexagons when it is subjected to uniform tension because a hexagonal system provides the greatest stress relief with the fewest cracks. Regular hexagons are rare, however, because cooling stresses in rocks are never completely uniform and the columns are generally bounded by curved cracks forming irregular-shaped polygons with variable numbers of sides. "Ideal" cooling conditions are never reached at the surface of a flow. However, as a progressive zone of cooling, solidification, and cracking proceeds from the surface of a flow into its interior, a point may be reached where the shrinkage forces may be uniform enough for earlier irregular jointing to give way to the formation of columnar joints. The importance of the homogeneity of the cooling rock explains why in the Portage Lake Volcanics well-developed columnar jointing is restricted to the very fine grained traps and has only a very slight tendency to form in the coarser grained porphyritic and ophitic rocks.

Columnar jointing can be seen in the Edwards Island Flow at all the localities mentioned previously. Other accessible areas where columnar jointing can be observed are where the trail from Rock Harbor to Mount Franklin crosses the flow and on the south slope of Ransom Hill just west of the trail from Daisy Farm Campground to Mount Ojibway.

The Greenstone Flow is the thickest flow on the island and holds up the most prominent ridge running the length of the island, Greenstone Ridge. The Greenstone Ridge Trail, which follows the ridge for nearly its full length, provides access to numerous scattered outcrops, but only at the far northeast end of the island near Blake Point is a reasonably complete cross section of the flow exposed. There the flow can be seen to consist of four divisions with approximate thicknesses as indicated: a lower ophitic zone (100 ft), a central pegmatitic zone (75 ft), an upper ophitic zone (175 ft), and an uppermost columnar-jointed trap (50 ft), for a total thickness of about 400 feet. The Greenstone Flow attains its greatest thickness in the central part of Isle Royale, where it is estimated to be nearly 800 feet thick.

The lower ophitic zone, which makes up the cliffs of the Palisades on the north side of Blake Point, is of particular interest as it is matched in coarseness only by the ophite of the Hill Point Flow; each has augite crystals as much as 2 centimetres in diameter. The pegmatite is best seen in the vicinity of the lighthouse on Blake Point. The columnar-jointed trap that forms the uppermost part of the Greenstone Flow is exposed on the string of small islands south of Merritt Lane, on Red Rock Point, and at a few additional isolated localities on the main island.

An excursion to Blake Point is also a botanical experience, as a rather unusual shrub fills a small ravine just west of the lighthouse—devilsclub. The entire plant, including its giant leaves, is profusely armed with sharp spines, and anyone who ventures to walk through a thick stand of these would appreciate the plant's specific name—*Oplopanax horridum* (fig. 19). It presence on Isle

DEVILSCLUB. (Fig. 19)

Royale is truly remarkable, for it appears nowhere else in the United States east of the Rocky Mountains; at Isle Royale it is found only near Blake Point, on Passage Island, and on a few of the small islands in the Rock Harbor area.

Pyroclastic and sedimentary rocks are exposed in only a few localities in the Rock Harbor—Tobin Harbor area. Pyroclastic rock can be seen on the north shore of Tobin Harbor opposite Newman Island and on Porter Island above (southeast of) the columnar-jointed upper part of the Greenstone Flow. The rock is a tuff-breccia, composed of angular fragments of volcanic rock cemented together by ash (fig. 20). The ash was in a hot and partly plastic state when deposited and was then fused into a coherent mass upon cooling.

An extensive wave-swept outcrop of conglomerate is exposed at the southwest end of Mott Island, and a few other less accessible outcrops are scattered along the rest of the chain of islands south of Rock Harbor (fig. 21). Volcanic rocks of many varieties can be identified as pebbles in the conglomerate. Many of them are of rhyolite or other types different from those that form the bulk of the Portage Lake volcanic sequence and must have been derived elsewhere.

PYROCLASTIC ROCK—note fragmental character. (Fig. 20)

CONGLOMERATE ON MOTT ISLAND (southwest end of island) showing pebbly surface. (Fig. 21)

Washington Harbor Area

The southwest end of Isle Royale is mantled with a much thicker blanket of glacial till and other surficial materials than the northeast end, and consequently, outcrops are common only along shorelines with exposure to storm waves. In the Rock Harbor—Tobin Harbor area, we were discussing rocks that make up the upper part of the volcanic sequence, that is rocks from the Greenstone Flow upward in the sequence. In the Washington Harbor area, most exposures are of the sequence downward from the Greenstone Flow (fig. 22).

Most of the rocks exposed in the Washington Harbor area are of ophitic basalt, as for example on the series of small promontories from North Gap northward. Exceptionally coarse ophite can be seen in the Hill Point Flow at the entrance to Huginnin Cove, accessible via the Huginnin Cove Trail as well as by boat. Some large loose slabs of the coarse porphyrite of the Huginnin Flow lie on the west shore of the cove (fig. 23), and excellent exposures of that flow can be seen on the north

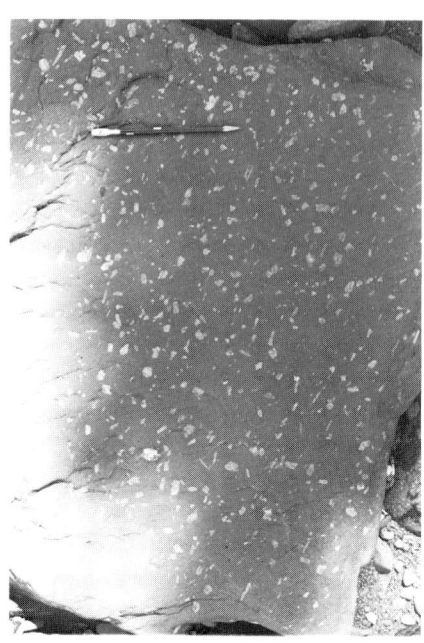

HUGINNIN FLOW PORPHYRITE—sample of loose slab from Huginnin Cove. (Fig. 23)

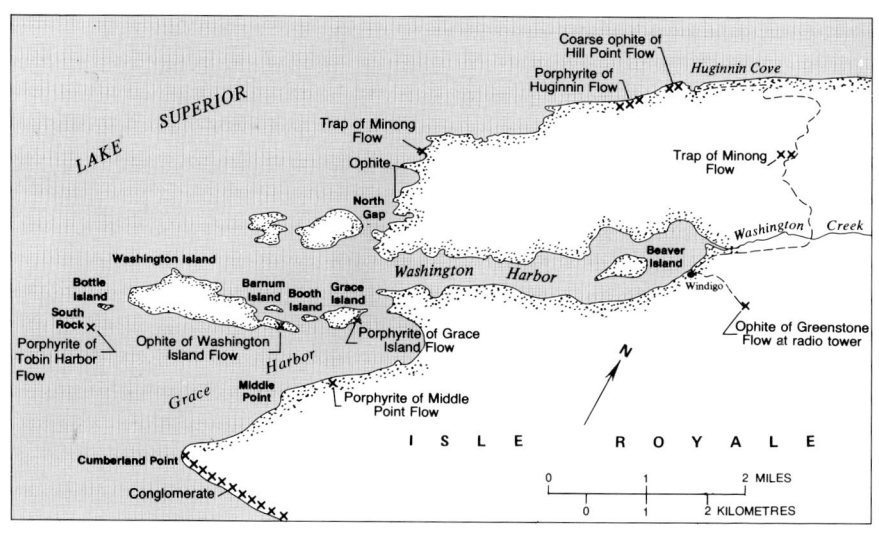

WASHINGTON HARBOR AREA—selected outcrops of various rock types. (Fig. 22)

shore of the island about half a mile southwest of the cove. Working upward in the volcanic sequence, trap of the Minong Flow is exposed along the Huginnin Cove Trail and forms a small promontory near the northwesternmost corner of the main island.

Outcrops of the coarse porphyrite of the Grace Island Flow are found on Grace Island just south of the campground, on Booth Island, and near both ends of Washington Island. Beach cobbles of rock from this flow are readily identifiable (fig. 24).

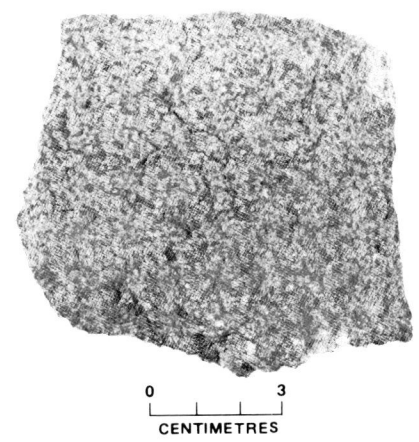

WASHINGTON ISLAND FLOW OPHITE showing distinctive dark splotches of chlorite. (Fig. 25)

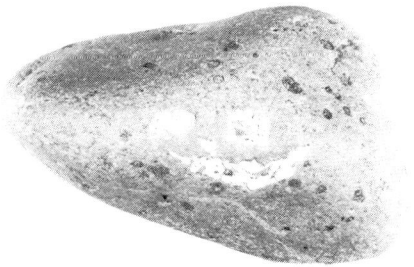

GRACE ISLAND FLOW PORPHYRITE— sample of beach cobble from Grace Island. (Fig. 24)

The rock of the Washington Island Flow is difficult to describe but distinctive when seen; perhaps A. C. Lane's application of the term "glomeroporphyritic ophite" suggests the problem. The rock is definitely an ophite, but the ophitic texture is obscure; instead the rock is characterized by a speckled appearance due to the rather uniform distribution of greenish splotches of chlorite (fig. 25). The Washington Island Flow is well exposed on the south side of Washington Island where it overlies the Greenstone Flow and caps the highest ridge.

South Rock, just southwest of Washington Island, is a herring gull rookery. It is also one of only two small outcrops of fine-grained porphyrite that are reasonably accessible near the west end of Isle Royale. It is believed to be part of the Tobin Harbor Flow. Another very small outcrop of porphyrite of the Middle Point Flow can be seen on the south shore of Grace Harbor about half a mile northeast of Middle Point.

AN ANCIENT ALLUVIAL FAN— THE COPPER HARBOR CONGLOMERATE

The shoreline of the southwestern part of Isle Royale from Cumberland Point to Malone Bay lacks the dark, forbidding cliffs typical of most of the island. There in many places rusty-red rocks slope rather gently down to the water (fig. 26). Pebble beaches are moderately abundant, whereas they are quite rare elsewhere (fig. 27). It is readily apparent that the entire appearance of the island is different here, and the reason is geologic. This part of the island is underlain by sedimentary rocks of the Copper Harbor Conglomerate.

COPPER HARBOR CONGLOMERATE sloping gently southeast into Lake Superior near Attwood Beach. (Fig. 26)

SANDSTONE PEBBLE BEACH west of Little Boat Harbor. National Park Service boat *D. J. Tobin* provided transportation for geologic studies. (Fig. 27)

The name of this formation is somewhat misleading because on Isle Royale it includes an appreciable amount of sandstone as well as conglomerate. On the Keweenaw Peninsula where the formation was originally defined, however, conglomerate is overwhelmingly predominant and hence the name. We shall see that within this formation on Isle Royale there is a definite pattern to the distribution of the sandstone and conglomerate that helps us to interpret where and how the original sediments were laid down.

The sedimentary rocks, especially the sandstones, commonly exhibit a well-developed parting parallel to the original layering or bedding in the rock (fig. 28). It is this parting, along with the lakeward slope of the strata, that permits storm waves to remove thin slabs of rock so as to produce a shore that slopes into the water. However, where the parting slopes away from rather than toward the water, as on the north sides of islands and promontories, cliffs form. What is probably the highest sheer cliff on Isle Royale is cut into conglomerate on the north side of Feldtmann Ridge (fig. 29).

The sedimentary rocks of the Copper Harbor Conglomerate are similar to those interbedded with the Portage Lake Volcanics. The pebbles and sand making up these rocks have predominantly been derived by erosion from volcanic rocks, many of which are similar to those exposed on the island, but many of which are more felsic (higher in silica, see p. 12). The felsic varieties can be recognized by their generally reddish tint. Agate pebbles that are locally abundant in the conglomerate are amygdules from volcanic rocks.

The rock sequence of the Copper Harbor Conglomerate is far from completely exposed on Isle Royale, as nearly half its total thickness is concealed beneath the mud and water, black spruce, white cedar, and bog mats of the Big Siskiwit Swamp. Nevertheless, enough can be deduced from the existing outcrops to reconstruct the environment in which the sediments were deposited and then to speculate upon the origin of the formation.

One clue comes from variations in thickness of the rock sequence. In spite of the fact that we cannot estimate the total thickness of the formation because the top lies beneath Lake Superior, the Copper Harbor Conglomerate has marker horizons that can be traced or projected for considerable distances, allowing us to calculate changes in thickness for equivalent parts of the formation. The thickness of the exposed part of the section increases rather steadily from Rainbow Point to the vicinity of Menagerie Island about 24 miles to the northeast (fig. 30). The Copper Harbor Con-

SLABBY SANDSTONE outcrops on Stone House Island. (Fig. 28)

CLIFF IN COPPER HARBOR CONGLOMERATE on north side of Feldtmann Ridge. (Fig. 29)

glomerate is a great wedge of sedimentary rock, and while we see it in only two dimensions, other clues will help to elucidate its nature more fully.

Another clue lies in the variable coarseness of the sedimentary debris that makes up the formation. Near the base of the formation, on Cumberland Point, the rock is a very coarse boulder conglomerate with boulders as much as 2 feet in diameter. Eastward the conglomerate decreases in coarseness, finally grading entirely into sandstone east of Malone Bay. Higher in the formation a similar change can be seen from boulder conglomerate at Rainbow Point and cobble and pebble conglomerates along the shore south of Feldtmann Ridge to sandstone on Houghton Point and on the chain of islands on the south side of Siskiwit Bay. Thus the size of the stones in the conglomerate decreases as the formation increases in thickness. This decrease in grain size suggests a corresponding increase in distance from the source of the sedimentary debris, as fine materials can be transported farther than coarse materials under the same conditions.

Other evidence regarding the location of the source of the sediments comes from structures preserved in the rocks themselves. Some of these structures indicate current direction and thus presumably the direction sand and gravel were carried. An example is a structure formed by the deflection of the water current around larger-than-average fragments (fig. 31). Analysis of such features in the Copper Harbor Conglomerate indicates that the direc-

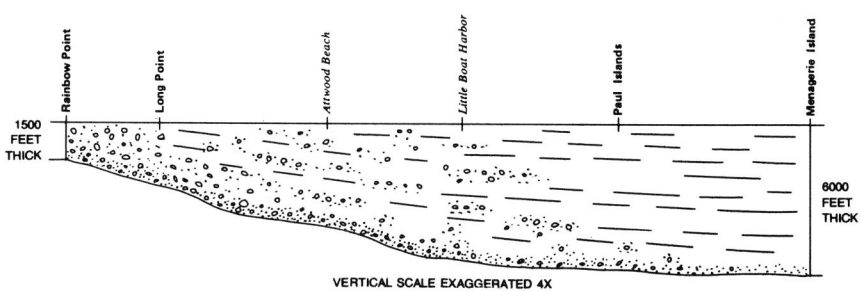

COPPER HARBOR CONGLOMERATE— section showing changes in thickness and in coarseness of grain size. (Fig. 30)

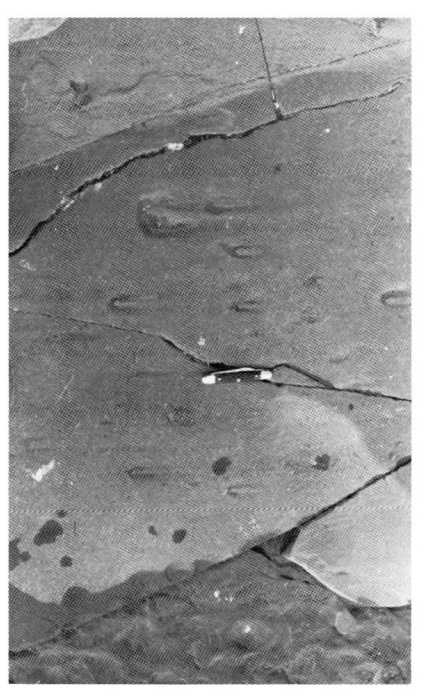

WATER-CURRENT FEATURE formed by flow deflected around shale fragments. Direction of current flow is toward the right. (Fig. 31)

tion of sediment transport was generally eastward (fig. 32).

One other clue helps to nail down the source area—the composition of the pebbles themselves. Felsites and some of the other volcanic rocks that occur as pebbles in the Copper Harbor Conglomerate are not found in flows in the exposed Portage Lake Volcanics on Isle Royale. However, such rock types are found in a somewhat older sequence of volcanic flows exposed on the north shore of Lake Superior in Minnesota, exactly in the direction indicated as the source by the current structures, the distribution of coarse conglomerate, and the thinning of the sedimentary wedge. As a well-known expert might say, happiness is having the right kind of rocks in the right place.

To sum up, the Copper Harbor Conglomerate is a vast wedge of sedimentary rock derived from volcanic rocks in what is now Minnesota and transported generally eastward to its present site on Isle Royale, with, as might be expected, the fine sediment particles being transported farther than the coarse ones. But what can we say about the nature of conditions at the time and place of deposition?

Most of the bedding structures as well as the lenticular form of zones of differing coarseness in the conglomerates are characteristic of deposits associated with the flowing water of streams and rivers rather than lakes or

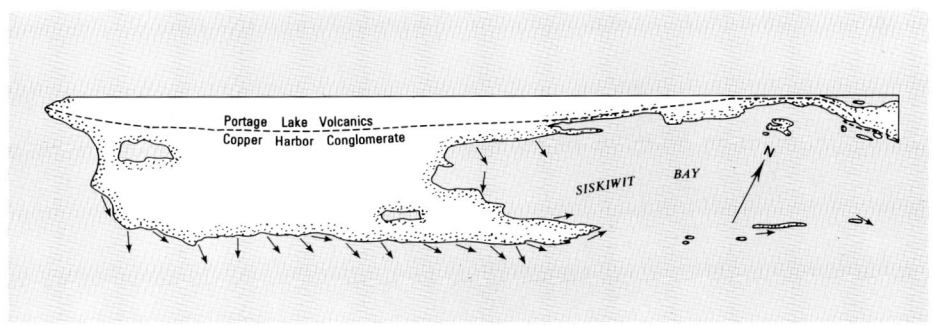

DIRECTIONS OF SEDIMENT TRANSPORT for the Copper Harbor Conglomerate. (Fig. 32)

oceans. Irregular scour-and-fill structure (fig. 33) is characteristic of sediment deposited by braided streams, ones flowing in dividing, reuniting, and constantly shifting channels resembling the strands of a braid (fig. 34).

SCOUR-AND-FILL STRUCTURE in pebbly sandstone, Chippewa Harbor area. (Fig. 33)

BRAIDED STREAM CHANNEL of the Muddy River, Alaska. Photograph by Bradford Washburn. (Fig. 34)

Some of the structures, such as ripple marks (fig. 35), do indicate deposition in standing bodies of water. But associated desiccation cracks and raindrop impressions (fig. 36) indicate that those bodies of water were temporary ones, such as might occur on river flood plains or mudflats.

RIPPLE MARKINGS on sandstone, west of Attwood Beach. Photograph by R. G. Wolff. (Fig. 35)

A

ANCIENT DESICCATION CRACKS (A) AND RAINDROP IMPRESSIONS (B) on siltstone at Isle Royale compared with similar features (C) on modern drying mud. Photograph of modern features by D. M. Baird. (Fig. 36). Figure continued on following page.

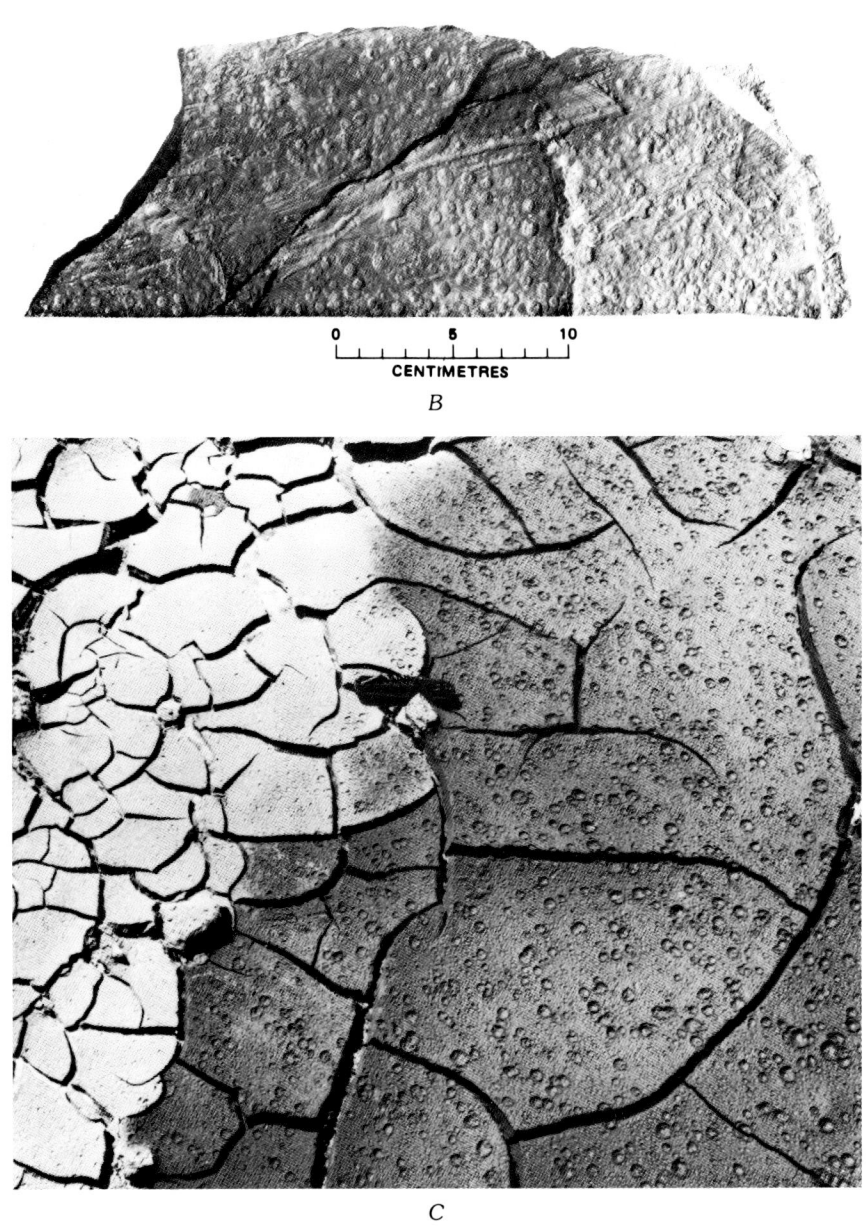

Figure 36—Continued.

The closest analogy we can find today of a sedimentary deposit with the features of the Copper Harbor Conglomerate are the massive alluvial fan deposits and associated finer grained playa deposits in arid regions of the

American West (fig. 37). The formation of such deposits requires very rapid, flash-flood runoff after rainstorms, rather than the continuously flowing water of permanent streams, and lack of vegetation is the key element permitting this to happen. This analogy does not necessarily mean that the Copper Harbor Conglomerate was deposited in a desert, for at the time the deposit was formed vegetation did not yet exist on land. In Precambrian time (see p. 36), before the appearance of a significant plant cover, erosion and runoff rates were high; floods were large, and coarse sediments were spread over piedmont areas more readily and rapidly than under modern conditions. Thus, the Copper Harbor Conglomerate on Isle Royale is a massive piedmont fan deposit, with associated flood plain or playa deposits, that was spread out over the Portage Lake Volcanics from a bordering highland area to the west.

TILTING AND BREAKING THE ROCKS

We have seen that Isle Royale consists of a thick series of flood basalt flows overlain by a great alluvial fan. Furthermore, the sequence has been tilted southeast toward the axis of the Lake Superior basin, and subsequent erosion has exposed the upturned edges of individual layers. The tilt or dip of individual layers varies from less than 10° to 55°. The dip is generally steeper on the north side of the island than on the south side and averages less than 20° for the island as a whole.

The only major distortion superimposed upon this generally parallel group of inclined beds is in the area between Siskiwit Lake and Rock Har-

ALLUVIAL FANS AND PLAYAS in Death Valley, Calif. From "Geology Illustrated" by John S. Shelton, W. H. Freeman and Company (copyright 1966). (Fig. 37)

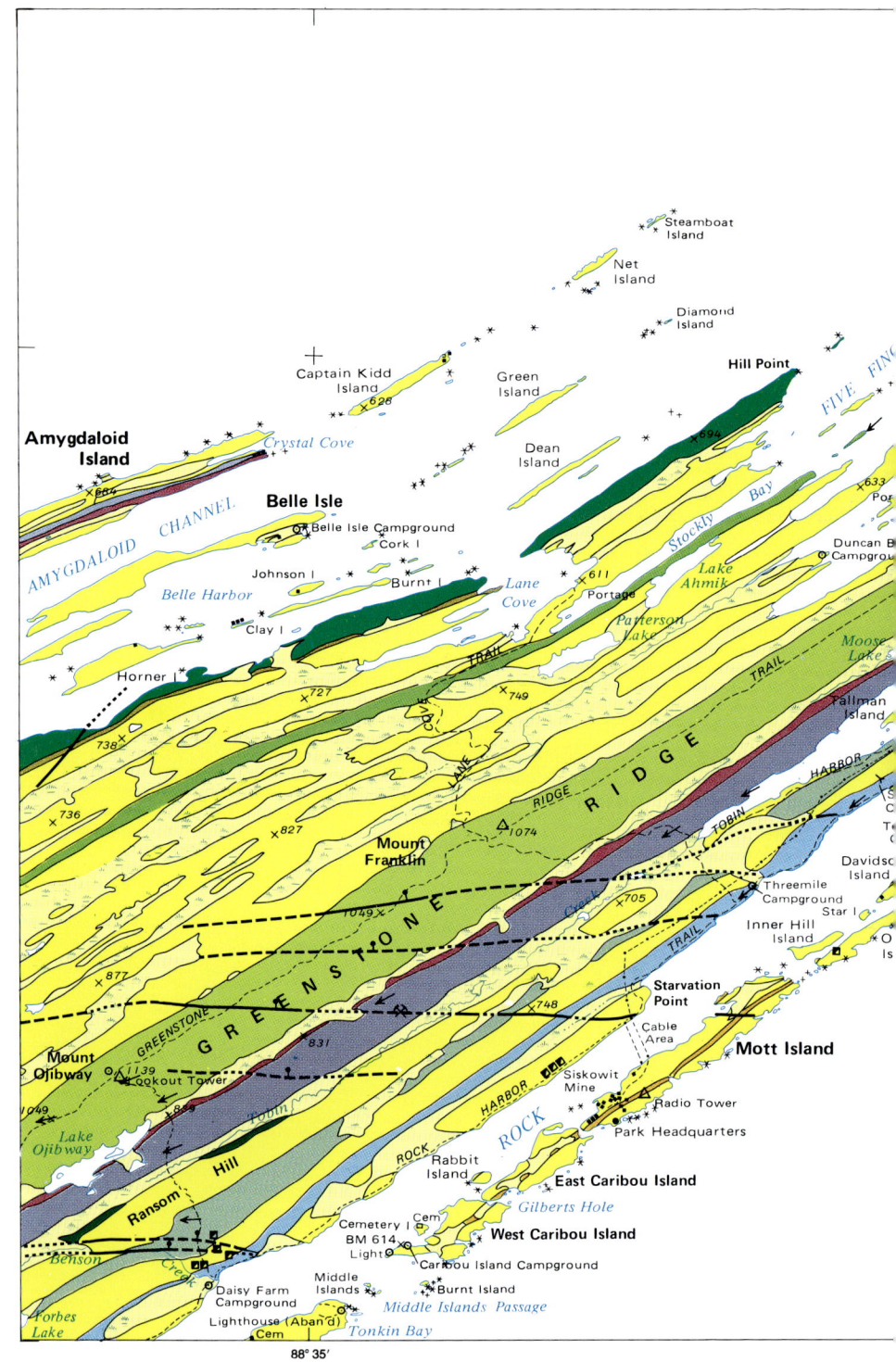

GEOLOGIC MAP OF THE NORTHEAST

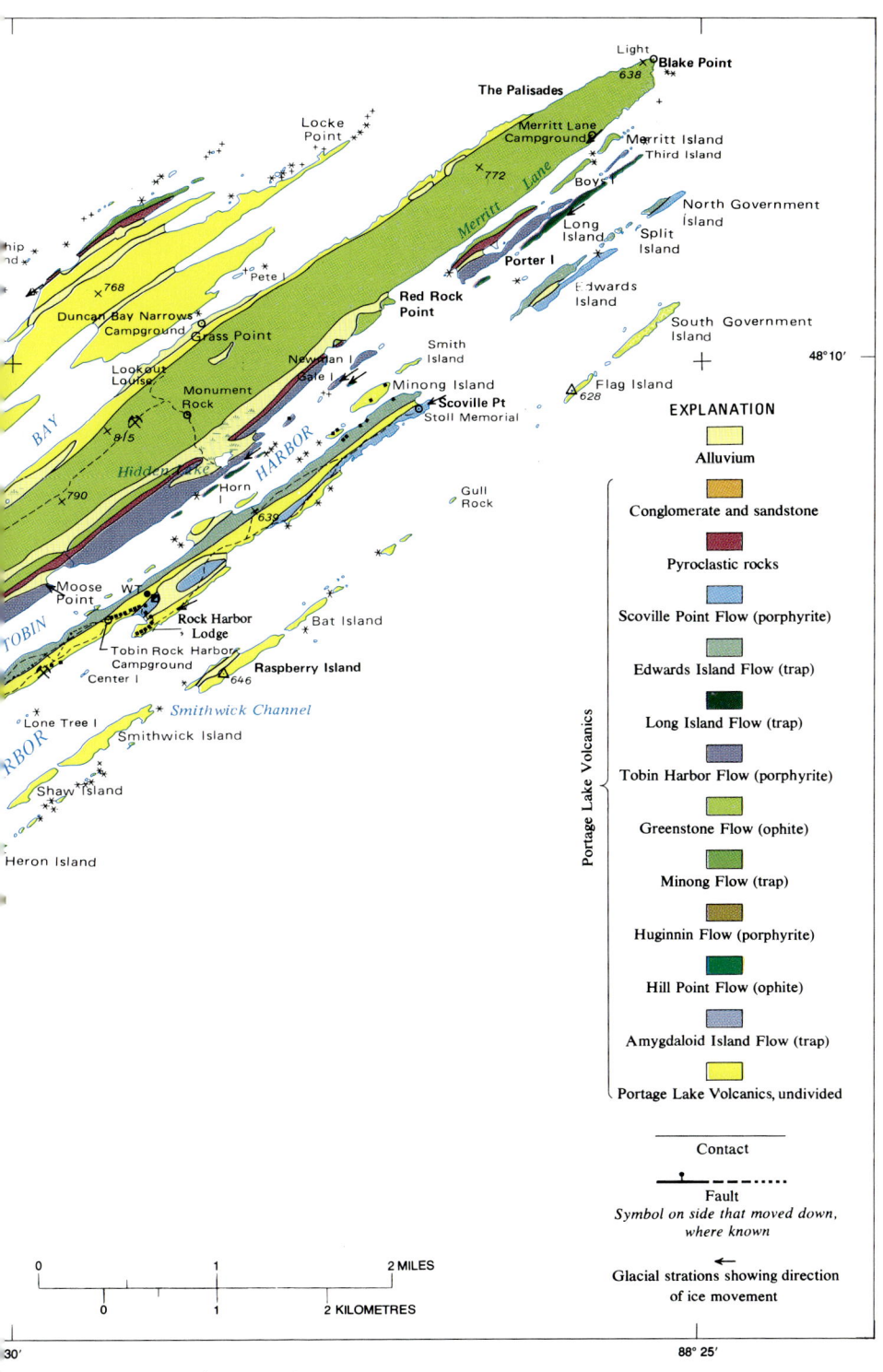

ISLE ROYALE (PLATE 1)

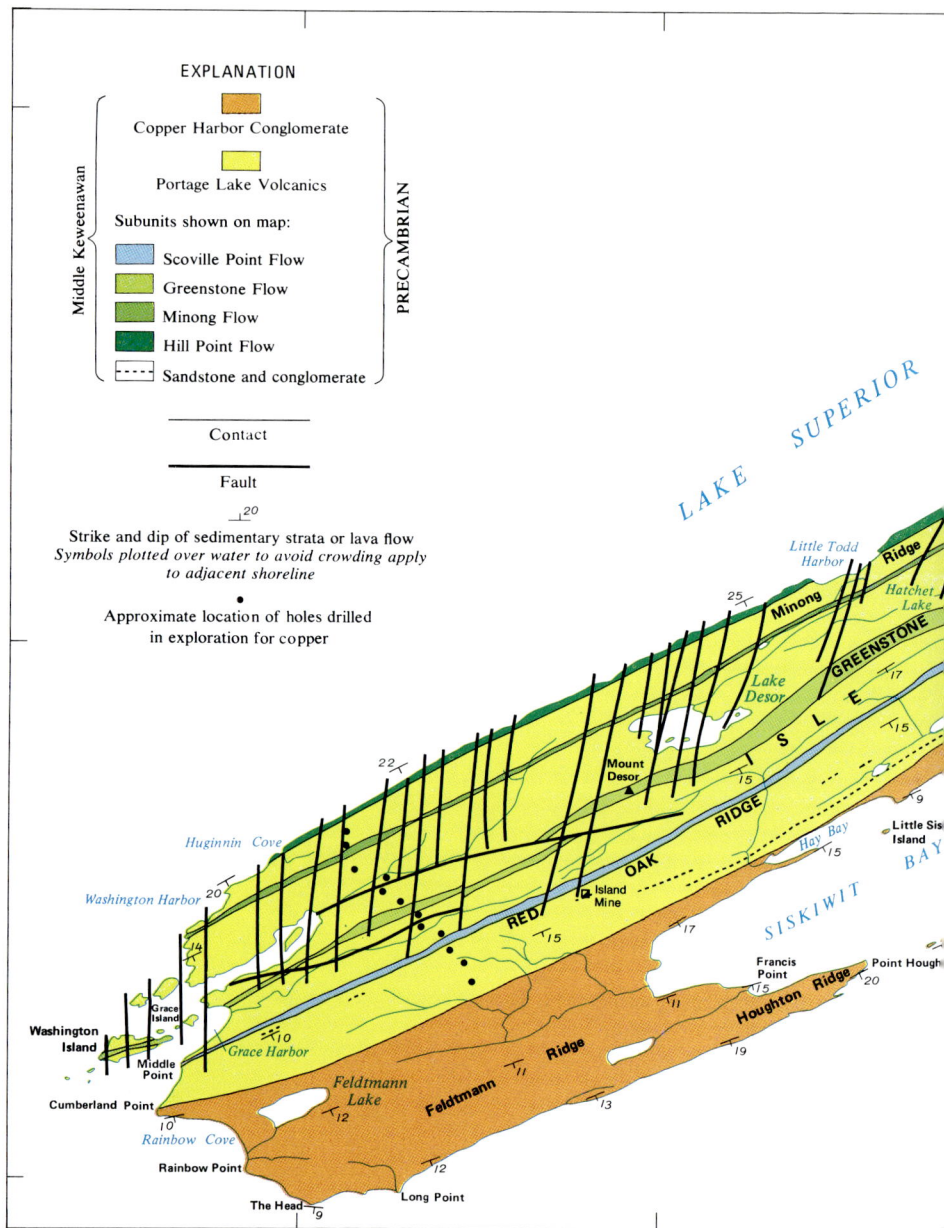

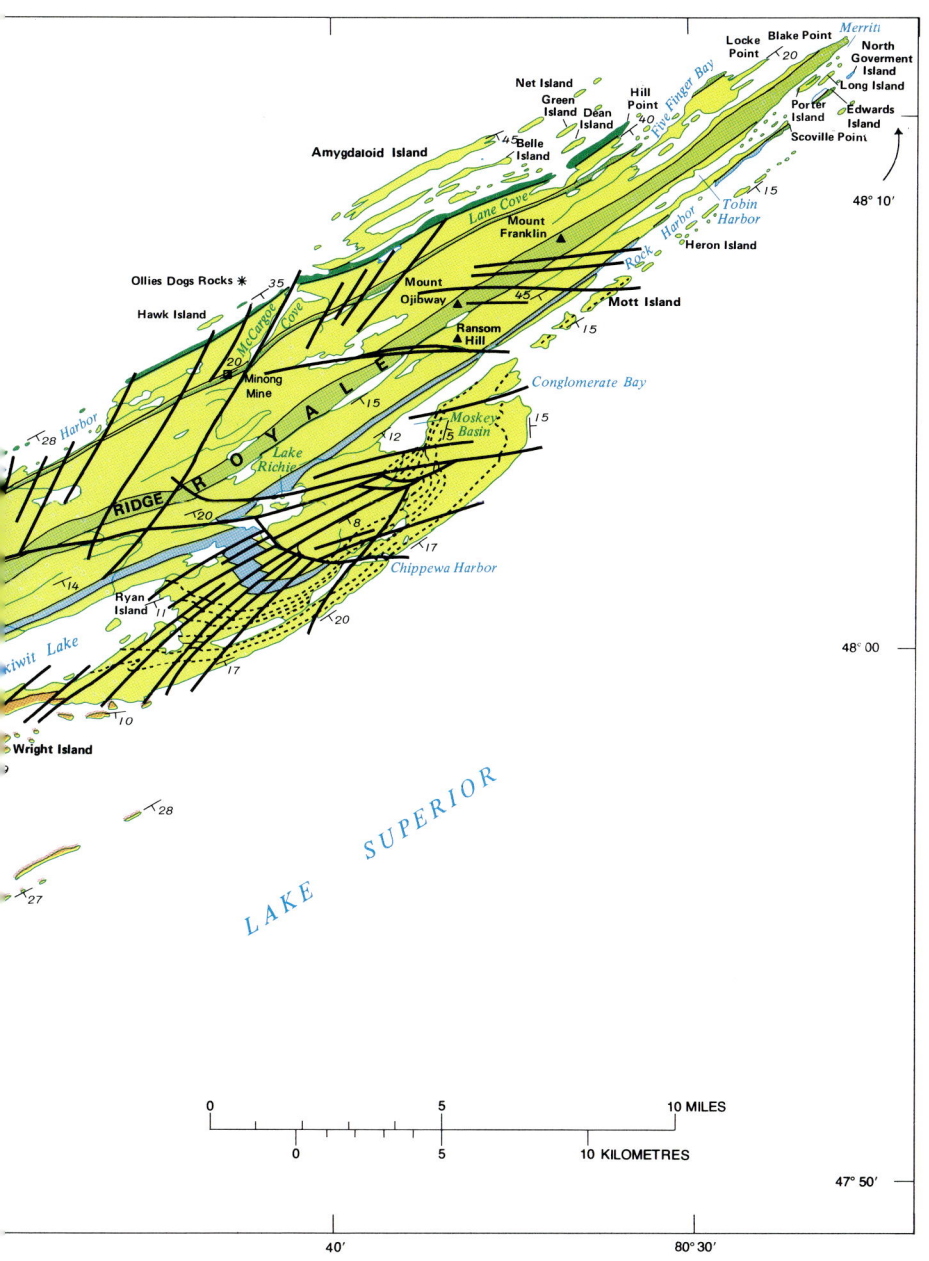

MAP OF ISLE ROYALE (PLATE 2)

bor, where the upper part of the Portage Lake Volcanics is warped around a node of apparently uplifted and fractured rocks (pl. 2). In appearance the distortion is something like a knot in an otherwise straight-grained piece of wood. The area appears to have been domed upward slightly to create this structure, but the specific cause is unknown.

Groups or sets of nearly parallel fractures cutting across the linear ridge system are also superimposed upon the tilted strata. Erosion along these fractures sometimes creates conspicuous ravines, and one set is clearly visible on the shaded-relief map (see title page). This set trends northward at the west end of Isle Royale and changes progressively eastward to trend about 30° east of north at McCargoe Cove. Minor fault displacement has occurred along some of the fractures in this set, but the amount of movement was generally small. A second set of fractures, trending east-west and not as apparent on the shaded-relief map, commonly shows somewhat greater fault displacement but still no more than a few hundred feet. A third set of curved fractures is present in the structurally disturbed area between Siskiwit Lake and Rock Harbor. All the fractures were probably formed as the beds broke up during tilting and warping of the layered rock sequence.

ISLE ROYALE— A SMALL PIECE OF THE BIG PUZZLE

The orientation of the beds on the Keweenaw Peninsula is a mirror image of that on Isle Royale. The bedrock sequence, similar to that on Isle Royale, generally dips northwestward toward the axis of the Lake Superior basin. Recognition of these similarities permitted C. T. Jackson to state as early as 1849 that "this island has the same geological character as Keweenaw Point, and is of the *same geological age*." He and his colleagues, J. W. Foster and J. D. Whitney, furthermore considered the rock sequences in the two areas to be connected beneath Lake Superior and thus interpreted the lake as occupying a structural basin, a syncline, as well as a topographic depression (fig. 38).

The later work of A. C. Lane not only reinforced this interpretation but also demonstrated that some individual lava flows or groups of flows, as well as some sedimentary rock units, on Isle

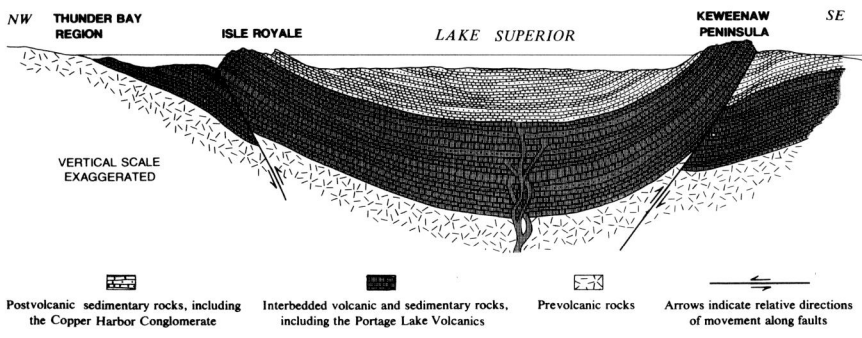

Royale were the same as those on the Keweenaw Peninsula (fig. 39). The continuity of the Greenstone Flow across the Lake Superior basin is most convincing, and its name is used in both localities; specific correlation of other flows is less certain, and separate names are applied to them on opposite sides of the lake.

The Portage Lake Volcanics on Isle Royale and on the Keweenaw Peninsula represent only the upper part of the total Keweenawan volcanic sequence in the Lake Superior region. The lower part of the sequence is exposed in the so-called South Trap Range of westernmost Michigan and along the north shore of Lake Superior in Minnesota and Ontario. Different formation names are given to these rocks in different areas (fig. 39). The total volcanic sequence reflects a long period of volcanic activity with a complex history. Erosional debris derived from the lower part of the volcanic sequence occurs in the sedimentary rocks interbedded with the lava flows of the Portage Lake Volcanics. The interpretation is that the lower part of the sequence was being eroded at the margins of a basin while the later flows were still being erupted into the central part of the basin.

WHAT HAPPENED WHEN?

The geologic story of Isle Royale as presented up to this point has been largely a description of the geology as we see it now. But how did it get this way? And when? The search for answers to these questions involves considerable interpretation of geologic observations made on Isle Royale and in the Lake Superior region, together

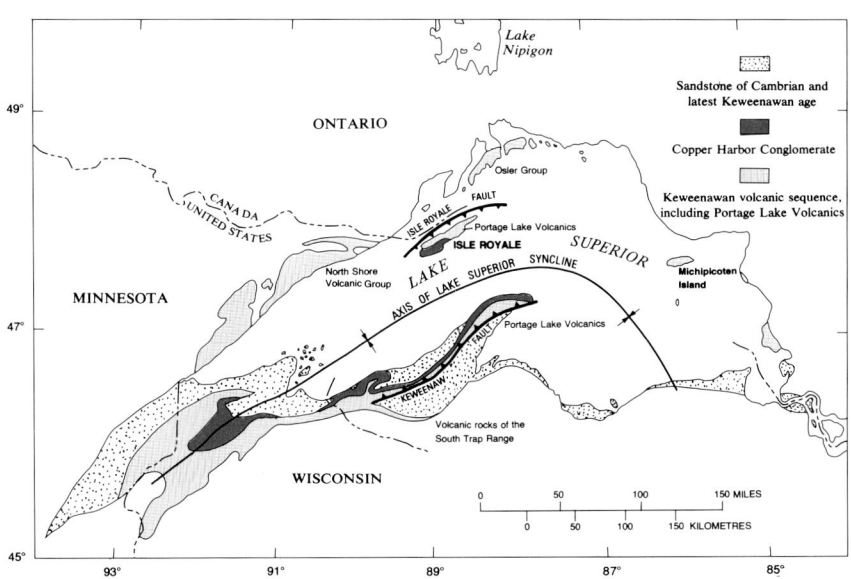

LAKE SUPERIOR REGION—distribution of selected rock units. (Fig. 39)

with numerous deductions based upon our accumulated geologic knowledge. Some parts of the geologic history can be deciphered in considerable detail, and other parts only incompletely because the geologic data are very spotty. The first step toward the answers is to develop a time framework within which to reconstruct events in the geologic past.

TIME AS A GEOLOGIC CONCEPT

A relative time scale, useful throughout the world, has been developed by dividing the last approximately 570 million years of geologic time into periods. The periods, which began with the Cambrian and end with the Quaternary, are recognized largely from the fossil record (fig. 40). Rocks of the Cambrian Period contain the earliest evidence of complex forms of life, which slowly evolved through the subsequent periods into the life of the modern world. The presence of distinctive fossils in various periods permits the correlation of rocks of similar age. However, the near absence of fossils in Precambrian rocks—those older than 570 million years—severely limits the use of fossils for the relative dating of the Precambrian rocks, rocks which cover more than 80 percent of geologic time. The rocks on Isle Royale are unfossiliferous and of Precambrian age, and it is with this poorly classified segment of geologic time that we are mostly concerned.

Fortunately, a means for measuring geologic time without fossils has been developed from the long-known natural process of radioactive decay of certain elements. From such radiometric age determinations, many rocks can now be assigned ages in years, which for Precambrian rocks, in the near absence of fossils, are virtually the only means for long-range correlations.

The uppermost, or youngest, Precambrian rocks in the Lake Superior region consist of a thick sequence of volcanic and sedimentary rocks. This sequence has been named the

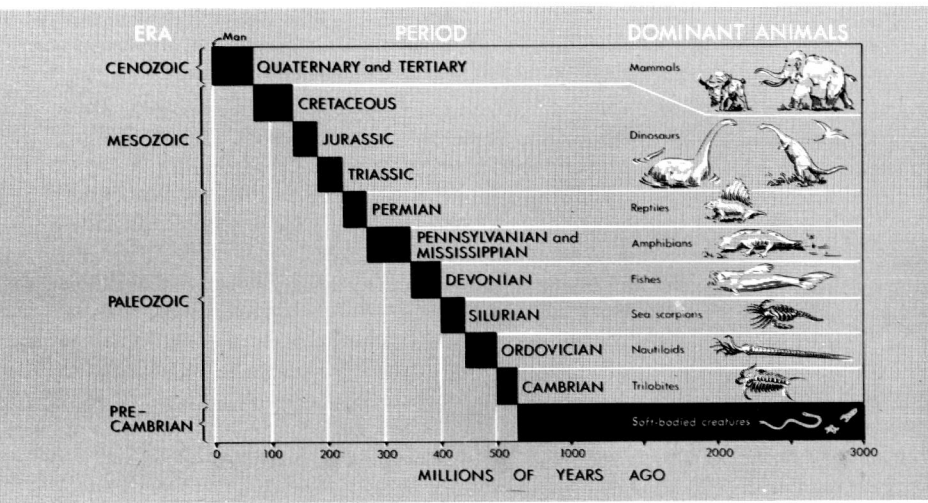

GEOLOGIC TIME CHART showing major divisions. (Fig. 40)

Keweenawan Supergroup because many of the formations that make it up were first described on the Keweenaw Peninsula and adjacent parts of Michigan and Wisconsin. The period of time during which the Keweenawan rocks were deposited can be referred to informally as Keweenawan time, recognizing that the term has usefulness only in the Lake Superior region where the Keweenawan rocks themselves occur. The Keweenawan Supergroup has been informally divided into lower, middle, and upper parts; formations exposed on Isle Royale, the Portage Lake Volcanics and the Copper Harbor Conglomerate, are assigned to the middle Keweenawan (fig. 41). We do not have sufficient data to determine the bounds of Keweenawan time, but radiometric age determinations indicate ages in the general range of 1,120–1,140 million years ago for the Keweenawan volcanic rocks. It is from this point in the enormously distant past that we begin to unravel the geologic history of Isle Royale.

WHAT WAS THERE BEFORE—THE PRE-KEWEENAWAN

Rocks of Keweenawan age, as old as they are, rest upon still older Precambrian rocks that are strikingly different. The older, or basement, rocks are of many varieties, ranging from granite and other intrusive igneous rocks to volcanic and sedimentary rocks, and cover a wide range in age. They usually are both more highly metamorphosed, or altered from their original state, and more strongly deformed than rocks of the Keweenawan sequence. Basement rocks are not exposed on Isle Royale, but they are sufficiently well exposed in areas both north and south of Lake

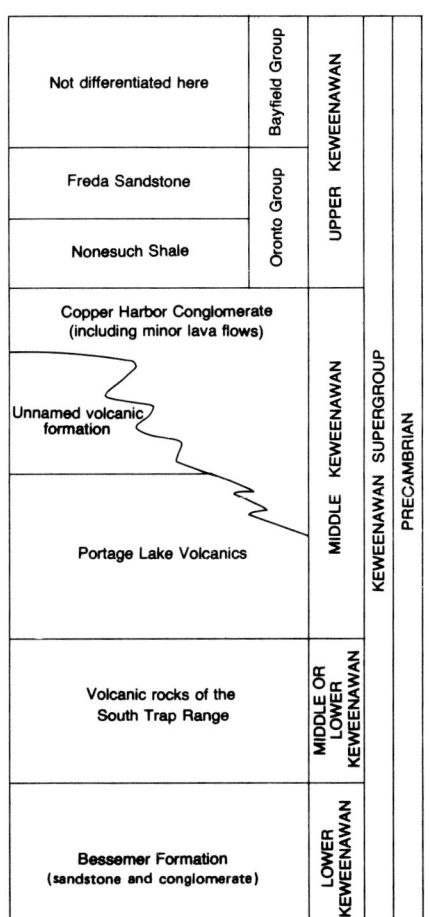

KEWEENAWAN SUPERGROUP—classification in Michigan. (Fig. 41)

Superior for their nature to be studied and their relationship to the overlying Keweenawan sequence to be determined.

The history of the pre-Keweenawan rocks is extremely complex, as might be expected from the wide range in rocks and age; however, prior to the time the Keweenawan rocks were deposited, much of what is now the Lake Superior region had been reduced to an area of fairly low relief, and shallow seas covered large parts of it. A thick sequence of marine sedimentary and volcanic rocks was then formed in those seas. In addition to impure sandstone and shale, chemically precipitated sediments were deposited, including dolomite and the iron-rich rocks that were later to make the Lake Superior district world famous as a producer of iron ore. The material forming the pre-Keweenawan volcanic rocks was erupted under water, where quick quenching of the lava caused it to congeal in irregular-shaped globular or ellipsoidal masses that accumulated one upon the other at the bottom of the sea (fig. 42). These volcanic rocks are thus quite different from the uniformly layered widespread sequence of Keweenawan flood basalts.

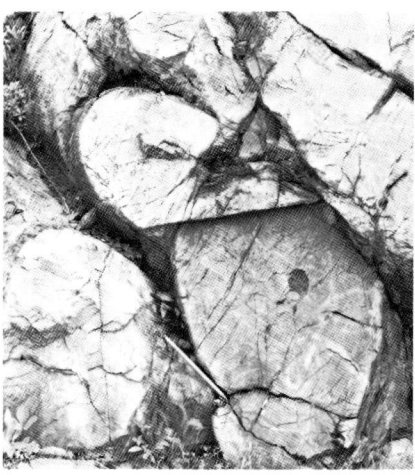

ELLIPSOIDAL STRUCTURES indicative of extrusion of lava under water, found in Precambrian volcanic rock from northern Michigan. (Fig. 42).

Toward the close of pre-Keweenawan time, the rocks present were folded, deformed, and eroded, quite drastically in some areas and only slightly in others, and so the material forming the overlying Keweenawan volcanic and sedimentary rocks was laid down on the edges of the older beds in some localities.

THE ROCKS OF ISLE ROYALE— THE KEWEENAWAN

The lowermost Keweenawan strata, which are not present on Isle Royale, consist of a relatively thin sequence of sedimentary rocks, chiefly conglomerate and sandstone, that lie on the pre-Keweenawan rocks. They have been interpreted as shallow-water, near-shore deposits and are exposed in only a few localities along the perimeter of the area of Keweenawan outcrops around the margins of the Lake Superior basin. The lowermost few of the Keweenawan lava flows above these sedimentary rocks contain ellipsoidal structures typical of those formed when lava is erupted into a body of water. Such structures are found only near the base of the lava sequence, however, and we conclude that the first few of the flood basalt flows were erupted into a shallow body of water, which was soon filled or disappeared for one reason or another, and that all later flows were on virtually dry land. Furthermore, as mentioned earlier, the sedimentary rocks interbedded with the lava flows show characteristics typical of those formed in streams rather than in lakes or oceans.

The source of the lavas appears to have been along the axis of an arc-shaped trough or system of fissures generally following the center of present-day Lake Superior. The system was actually much more extensive, however, extending southwest at least as far as Kansas and southeast into the southern peninsula of Michigan, a total distance of over 1,000 miles. Great volumes of lava were erupted from fissures near the axis of the trough and spread laterally and rapidly toward both margins of the trough, finally ponding and cooling in place to form vast sheets covering thousands of square miles. The Portage Lake Volcanics, as exposed on the Keweenaw Peninsula and Isle Royale, represent roughly the upper half of this volcanic pile (fig. 41).

The material forming the sedimentary rocks that are interbedded with the lava flows of the Portage Lake Volcanics and that are in the Copper Harbor Conglomerate and other Keweenawan formations above the lava sequence was transported by streams into the trough or basin from highlands around its margins. This evidence for inward flow of streams, contrasted with evidence that the lavas flowed toward the margins of the basin, shows that there were, at times, reversals of the prevailing slope over large areas and leads to the concept of a basin subsididing as it was being filled (fig. 43). The surfaces of the lava flows were horizontal or sloped gently toward the margins of the basin as long as filling by lava kept pace with downwarping. When extrusion of the lava was interrupted, however, continued downwarping produced inward slopes that permitted sedimentary debris to be swept into the basin. Finally, with the gradual demise of volcanic activity, continued subsidence permitted the accumulation of the Copper Harbor Conglomerate and younger Keweenawan deposits to form a thick sedimentary sequence above the volcanic rocks in the basin.

Sandstone of latest Keweenawan or earliest Cambrian age is exposed along much of the shore of the southeastern part of Lake Superior and extends westward to the Keweenaw Peninsula. The basin may have been largely filled

A. Lava erupts near the center of the basin and spreads laterally toward the margins to form a sequence of lava flows.

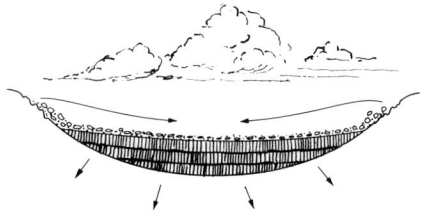

B. The basin subsides, and during a lull in volcanic activity gravels are swept into the basin and spread out over the uppermost lava flow.

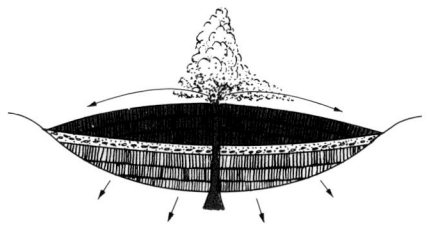

C. Volcanic activity resumes, and the cycle starts over again.

FLOOD BASALTS AND SEDIMENTS showing the process of interbedding. (Fig. 43)

by this sandstone, together with similar sandstone exposed in the southwestern part of the basin (fig. 39).

The gross synclinal form of the Keweenawan basin resulted from subsidence coincident with filling of the basin rather than later folding by squeezing. However, Keweenawan strata near the margins of the basin, as on the Keweenaw Peninsula and Isle Royale, were subsequently steepened by upward movement on major faults, the Keweenaw fault and the Isle Royale fault, thus accentuating the synclinal structure (fig. 38).

THE MISSING CHAPTERS

It is somewhat disconcerting that we can reconstruct so much of what happened during the Precambrian and so little of the geologic history of the Lake Superior region during the last 570 million years, including most of the Paleozoic, Mesozoic, and Cenozoic Eras. But this is the situation, for erosion has removed most of the critical evidence regarding that period of time. And, of course, much of the area is concealed by Lake Superior itself.

We do know, nevertheless, that the rocks of the Lake Superior region constitute the southernmost part of an area known as the Canadian Shield—an area of predominantly Precambrian rock and one that for the most part has been relatively stable geologically since the Precambrian (fig. 44). Beginning in Late Cambrian time, however, the so-called Michigan Basin, centered on what is now the southern peninsula of Michigan, began to subside, and a thick sequence of Paleozoic rock accumulated in the invading sea. Lower Paleozoic rocks at the margin of this basin extend up onto the shield near the southeast shore of Lake Superior. A few remnants of lower Paleozoic dolomite occur as far west as the base of the Keweenaw Peninsula; they are very thin and were probably deposited very close to the outer margin of the Michigan Basin.

We also know that lower Paleozoic rocks are present in a fairly extensive area centered on Hudson Bay. They include various types of rocks of shallow marine origin and in the vicinity of James Bay are locally overlain by nonmarine Cretaceous sedimentary rocks that include coal. Remnants of lower Paleozoic strata at a few other scattered localities on the Canadian Shield indicate that such strata were once more extensive than now, but how extensive is unknown. It is possible that they covered the entire shield, but we have no direct evidence that such rocks were ever present in the vicinity of Isle Royale.

Much of the Canadian Shield probably had been eroded to a relatively flat surface prior to the deposition of the lower Paleozoic strata. It remained so during the deposition and the subsequent erosion of much of those strata, right up to the Pleistocene Epoch and the onset of glaciation. Sometime after the deposition of the Jacobsville Sandstone of Precambrian or Cambrian age, however, movement on the Keweenaw fault thrust the Portage Lake Volcanics up over the Jacobsville Sandstone on the Keweenaw Peninsula. Presumably at about the same time, thrusting on the Isle Royale fault left the volcanic rocks of Isle Royale standing as an elongate ridge surrounded by softer sedimentary rocks. With this last disturbance the present geologic structure of the

THE CANADIAN SHIELD (stippled area)—predominantly Precambrian rocks. (Fig. 44)

Lake Superior region was established, and little has happened since except for erosion by both water and ice. A broad river valley probably developed in the present site of Lake Superior, as well as in the sites of the other Great Lakes, for all are located in belts of rocks less resistant to erosion than their bounding areas. The gross features of the preglacial landscape probably exerted a major influence on the direction of ice movement as successive lobes of glacial ice invaded the region during the Pleistocene and moved through the ancestral "Lake Superior valley."

THE GLACIERS TAKE OVER

The concept of widespread glaciation during a so-called ice age evolved over a relatively short period of time during the first half of the 19th century. It was first clearly expounded by Louis Agassiz beginning in 1837 but met with considerable skepticism for many years. Interestingly enough, Agassiz himself visited the Lake Superior region in 1848 and remarked that nowhere had he seen evidence of glacial phenomena in greater clarity than on the shores of Lake Superior. He traversed the north shore of the lake as far west as the Hudson Bay Co. post at Fort William, opposite Isle Royale on the Canadian shore, but he was unable to visit the island because of stormy weather. The next year Edouard Desor, who had been Agassiz's secretary and coworker for the previous 10 years, became associated with Foster and Whitney in their studies of the Lake Superior region and was the first to describe glacial features on Isle Royale. Mount Desor, the highest summit on Isle Royale, and Lake Desor, the second largest lake on the island, bear his name.

Beginning about 2 or 3 million years ago, the earth underwent a series of cyclic climatic changes, with at least four episodes cold or moist enough to permit the formation of extensive ice sheets. In North America, ice accumulated at several separate centers, spread out, and coalesced to cover a large part of the continent during each of these four episodes that together make up the glacial or Pleistocene Epoch (fig. 45).

Isle Royale was overridden by glacial ice during each of the four major glaciations, and each successive glaciation virtually obliterated all direct evidence of the preceding glaciations on the island. In the waning phase of the last major glaciation, the Wisconsin Glaciation, the frontal ice margin retreated northward from at least the greater part of the Lake Superior basin, then readvanced into the basin about 12,000 years ago (fig. 46). We can attribute the final carving of the Isle Royale landforms to this last advance, although all the earlier glaciations played their part.

Glacial Effects On The Land

The effects of glacial erosion were strongly dependent upon the nature of the bedrock. Over much of the Canadian Shield, where the bedrock consists of granite and other resistant Precambrian rocks, the ice sheets probably did little more than remove surficial materials, weathered rock, and minor amounts of bedrock. However, where the bedrock consisted of softer, sedimentary rocks, as along the axes of each of the Great Lakes, the ice sheet deepened the preglacial lowlands to form the present lake basins. During the excavation of the Lake Superior basin, Isle Royale remained a ridge because its volcanic rocks resisted erosion.

PLEISTOCENE ICE SHEETS showing the main centers of accumulation and the maximum extent of the Pleistocene ice sheets. (Fig. 45)

Plucking and Abrasion

Glacial erosion of bedrock includes two distinct processes: plucking and abrasion. Glacial plucking or quarrying is the lifting out and removal of fragments of bedrock by the ice, whereas glacial abrasion involves grinding and polishing of the bedrock surface by rock fragments imbedded in the base of the moving ice sheet. Plucking is by far the more effective in altering the landscape.

Glacial plucking is controlled chiefly by bedrock structures, particularly preexisting joints or other fractures. The sedimentary rocks and amygdaloidal zones at the base and top of flows are less resistant to abrasion than the massive flow interiors, and the sedimentary rocks and thinner flows have closer spaced jointing than the massive interiors of the thicker flows. The alternation of massive, poorly jointed rock and highly jointed rock that guided the development of the preglacial topography by stream erosion also was the major control for glacial quarrying. The main role of glacial erosion on Isle Royale thus was to accentuate the asymmetry of the existing ridge-and-valley topography that had been formed earlier by stream erosion and to interrupt the preglacial stream channels with a series of small lake basins scooped out of the bedrock.

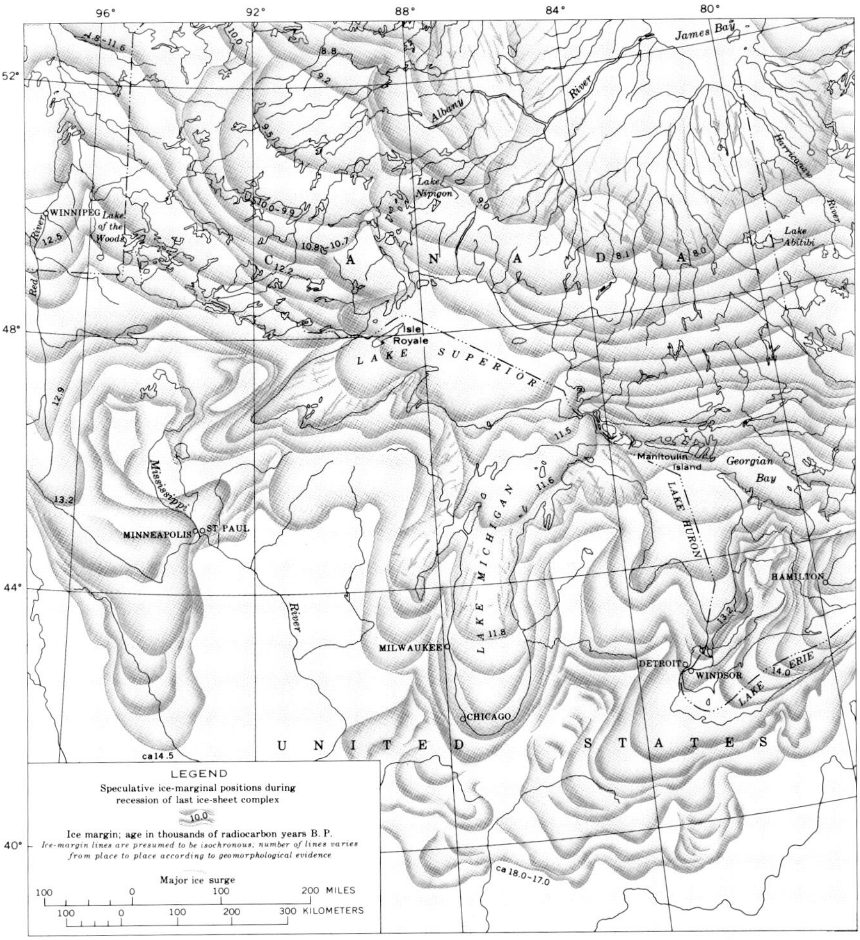

WISCONSIN ICE RETREAT from central North America. Note major ice surge, or readvance, in the Lake Superior basin (from V. K. Prest, 1969, Geological Survey of Canada Map 1257A). (Fig. 46)

Whereas glacial quarrying modified the gross topography of Isle Royale, glacial abrasion rounded, polished, and striated many rock outcrops (figs. 47, 48). The quarrying and scouring together were so effective, in fact, that deeply weathered rock, which we would expect to find after millions of years of weathering, is found only rarely on the island on the lee sides of a few outcrops where the weathered rock was protected from the action of moving ice.

GLACIAL STRIATIONS near Moskey Basin Campground. (Fig. 47)

GLACIALLY ROUNDED AND STRIATED OUTCROP near portage trail at Pickerel Cove. (Fig. 48)

Striations

Glacial striations on the bedrock not only indicate the former passage of glacial ice but also provide an indication of the general direction of movement of the ice that formed them (fig. 49). The striations indicate that the last ice movement on the east half of the island was toward the southwest, parallel to the bedrock ridge-and-valley topography. On the west half, striations indicate that the direction of movement was westward, crossing the ridges and valleys at an angle. The westerly direction is roughly perpendicular to a series of glacial moraines, or ridges of glacial till, that crosses the island near its west end; these moraines mark a pause in the retreat of the ice sheet. Southwesterly striations near the north side of the southwest end of the island presumably reflect the direction of the main advance down the Lake Superior basin prior to retreat to the position of the glacial moraines. Glacial striations are rarely preserved on the sedimentary rocks because the surfaces of those rocks weather and disintegrate relatively rapidly. Striations are best preserved on the denser, nonamygdaloidal volcanic rocks and can best be seen along shorelines where such rocks dip gently into Lake Superior or along shorelines of some of the larger inland lakes. Such locations are along the north shores of Tobin Harbor, Rock Harbor, Pickerel Cove, Lake Richie, Siskiwit Lake, and the west end of Mos-

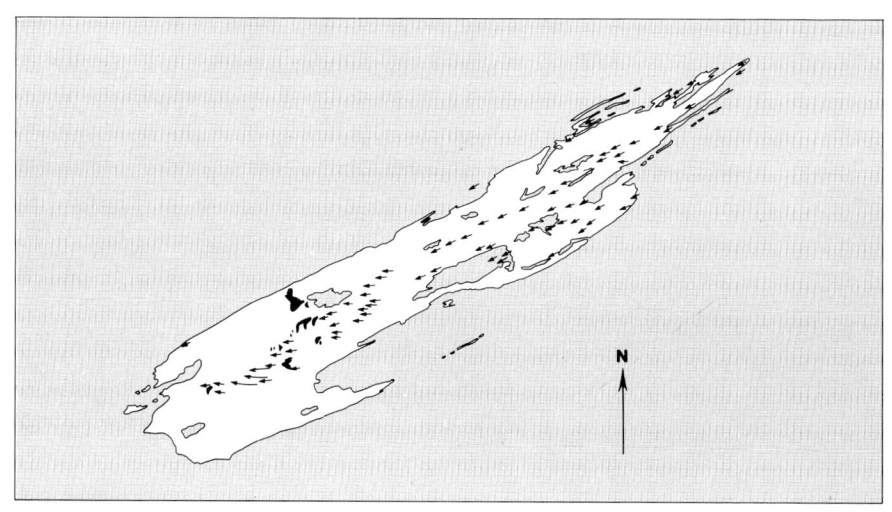

DIRECTION OF GLACIAL ICE MOVEMENT as indicated by striations and other glacial features. Dark areas are recessional moraines. (Fig. 49)

key Basin. Striations can also be seen along many of the inland trails, especially in places where the trail itself has channeled rainfall runoff so as to remove the soil cover and expose fresh bedrock.

Till

On the east half of Isle Royale, where the last ice advanced approximately parallel with the ridge-and-valley topography, surficial deposits other than a thin mantle of soil and rock debris are confined to valleys and smaller depressions, and conversely outcrops are relatively abundant except in the valleys; heterogeneous glacial deposits called till are not apparent. Scattered boulders of foreign rocks, called glacial erratics, do indicate the presence of glacially transported materials, but they are not overly abundant. Beach gravels along the present shoreline are predominantly derived from nearby bedrock, although they contain pebbles reworked from glacial materials. The most obviously foreign pebbles are of granitic rocks from the Canadian mainland and fossiliferous limestone and chert from the Hudson Bay area (fig. 50).

Near the west end of Siskiwit Lake, where the westward direction of ice movement crossed the trend of the ridges and valleys at an angle of about 25°, glacially deposited materials are more abundant; thin glacial till mantles many of the slopes, and outcrops decrease in size and number. Large erratics are numerous.

Till is abundant on the west end of the island, effectively concealing most of the bedrock and subduing the landforms (see title page). Large erratics also are abundant (fig. 51). Bedrock outcrops are mainly limited to small knobs and the steep north faces of some ridges. In many places, even these bedrock exposures have clearly been excavated from till by wave action at a time when Lake Superior was higher than now.

ERRATIC BOULDER of granite near crest of Feldtmann Ridge. (Fig. 51)

Drumlins.—West of Siskiwit Lake, where the till becomes more abundant, one can see deposits of till with distinctive shapes, certain of which are called drumlins. A drumlin is a streamlined hill or ridge of glacial till elongate in the direction of flow of the glacier that formed it. The drumlins on Isle Royale are of the crag-and-tail variety—linear ridges streamed out behind a bedrock knob (fig. 52). The drumlins vary in length from a few feet to nearly 2 miles; the larger ones can be seen on the shaded-relief map (see title page) and on aerial photographs as distinct east-west ridges superimposed upon the

FOSSIL SHELL in chert pebble. (Fig. 50)

45

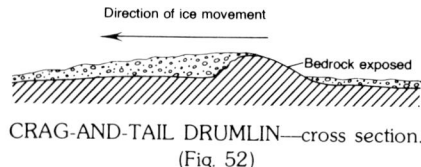

CRAG-AND-TAIL DRUMLIN—cross section. (Fig. 52)

bedrock-controlled, northeast-trending ridges (fig. 53). They are most abundant between the west end of Siskiwit Lake and a locale about 5 miles east of Cumberland Point. In an area west of Siskiwit Bay covering several square miles, the only bedrock exposed is at the heads of the drumlins. The elongation of the drumlins is useful to determine the direction of ice movement in areas where glacially striated bedrock is not exposed.

Recessional moraines.—A series of till-built recessional moraines that was formed during a pause in the retreat of the last ice sheet crosses the island in a general northerly direction west of Lake Desor. (See fig. 49). The form of these deposits is distinctive and can be generally characterized as an arc-shaped ridge roughly transverse to the direction of ice movement and not controlled by bedrock structure. The deposits define the former position of an irregular ice front controlled by topography, with tongues of ice extending into valleys or low areas in front of the main ice mass. (See fig. 59.) The ice front was not completely static, however, and continued to retreat by small increments. This intermittent retreat is

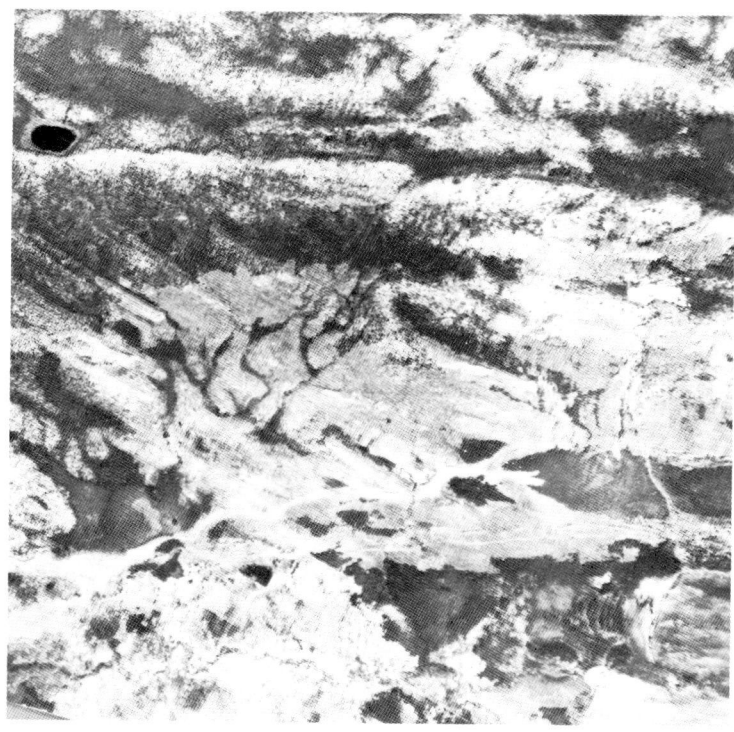

GLACIAL FEATURES FROM AIR including drumlins, abandoned shorelines, and other features of glacial origin west of Siskiwit Bay. (Fig. 53)

best illustrated by the succession of moraines on the south flank of Mount Desor that were constructed by the ice lobe that extended up the drainage basin of the Little Siskiwit River.

Glacial Effects On Lake Levels

Some uncertainty exists as to whether the last ice readvance completely filled the Lake Superior basin. But however far the ice did advance, when it retreated it left behind a sequence of lakes that progressively filled more and more of the basin. The retreating ice uncovered successively lower outlets, and thus the general trend of lake elevations was initially downward. Later lake levels were influenced by uplift of the final lake outlet as the earth's crust rose in response to the removal of the weight of the glacial ice, weight which had previously depressed the earth's elastic crust when the ice advanced southward.

In between rather rapid changes in lake levels caused by changes in outlets, the water remained at stable elevations long enough for waves to erode cliffs, build beaches, and construct other recognizable shoreline features. The history of the sequence of lakes that occupied the Lake Superior basin has been deduced largely from matching their abandoned shorelines around the basin. Such correlation of shorelines is complicated by the fact that as the shorelines are traced toward the northeast, in the direction of ice retreat, indi-

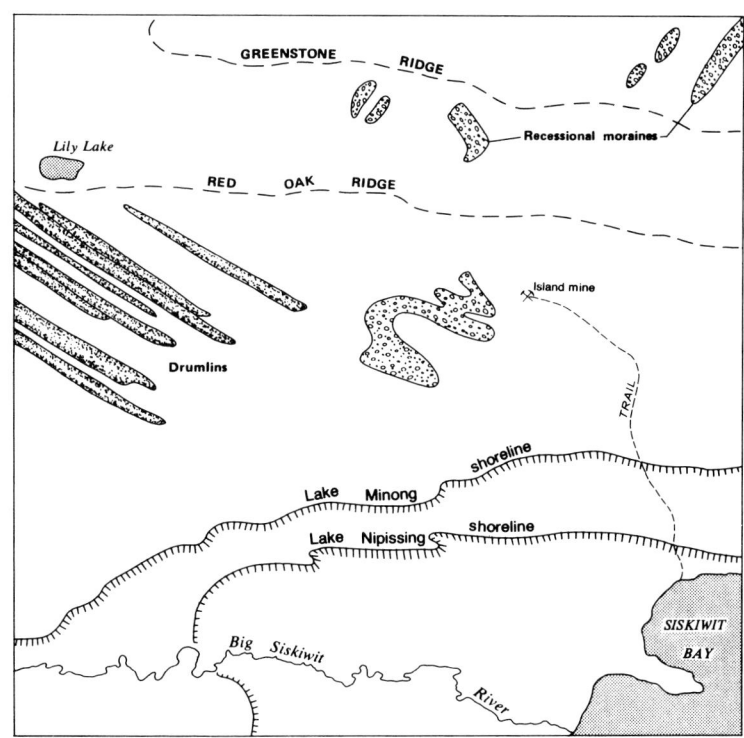

Figure 53.—Continued.

vidual shorelines gradually increase in elevation, owing to the rise of the land after the shorelines were formed (fig. 54). A specific example can be given by the shoreline of Lake Minong, one of the postglacial lakes. The elevation of its uplifted and warped shoreline on Isle Royale increases from about 80 feet above Lake Superior at the southwest end of the island to about 170 feet at the northeast end (fig. 55) a rise of over 2 feet per mile. Despite the complications introduced by tilting of the abandoned shorelines, through careful tracing of them, several distinct lake stages for the Lake Superior basin have been established and named. A shoreline diagram for the most important ones is shown in figure 56, and the chronology of lake-level changes is shown in figure 57.

An important fact to remember is that on Isle Royale all shorelines older than the present one have been uplifted to a varying degree, and thus their present elevations above sea level do not reflect the elevations of the lakes that formed them. The original elevations of individual lakes can be determined only where the shorelines remain undeformed, south of the region of crustal rebound or uplift (at the far left of figs. 54, 56). Note, for example, that Lake Minong, whose shorelines

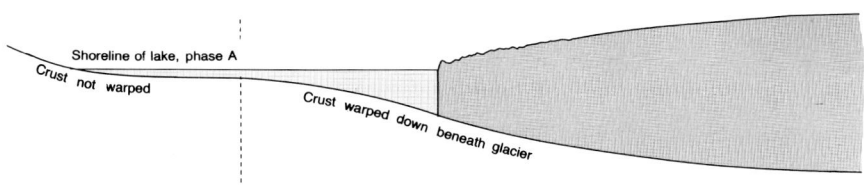

A. Glacial lake phase, A, has formed between the receding glacier margin and high ground to the left.

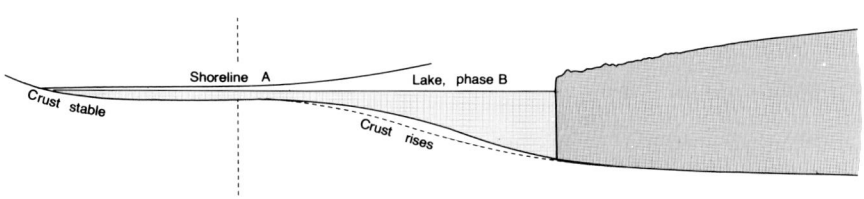

B. Uplift has affected the area to the right of the vertical dotted line, bending the shoreline of phase A to a higher position. A new lake phase, B, has formed at a lower level.

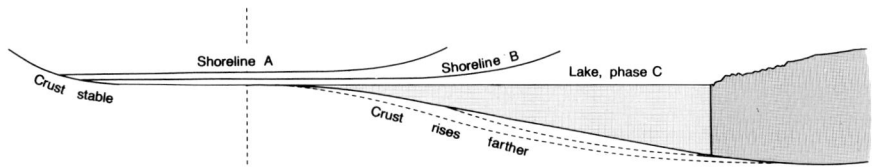

C. Renewed uplift has affected the area to the right of the vertical line, bending the shoreline of phase B and further warping the right-hand end of A. A third phase, C, has formed at a still lower level. The amount of warping is greatly exaggerated.

SHORELINE WARPING due to progressive uplift of the earth's crust as the glacier retreated (after Flint, 1971). (Fig. 54)

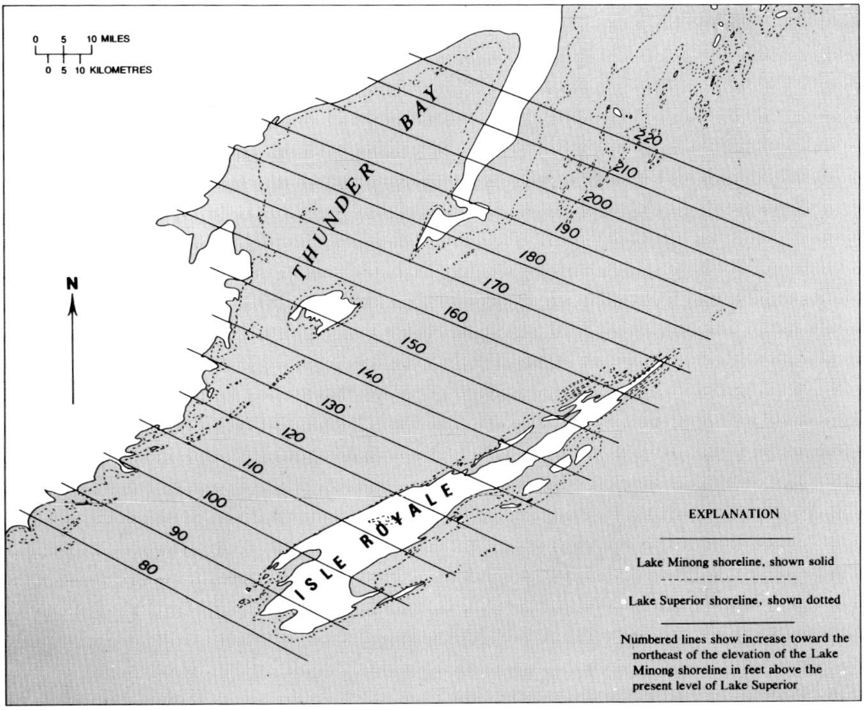

UPLIFT AND WARPING of the shoreline of postglacial Lake Minong. (Fig. 55)

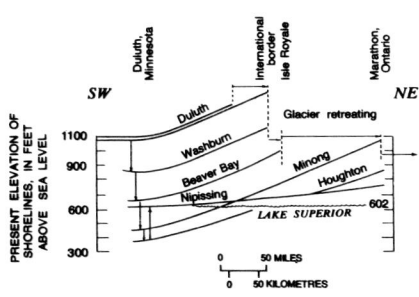

POSTGLACIAL LAKE SHORELINE STAGES in the Lake Superior Basin. Curvature is due to crustal rebound (uplift), progressively greater to the northeast. Horizontal arrows indicate progressive retreat of the ice margin; vertical arrows indicate direction of changes in lake elevations. (Fig. 56)

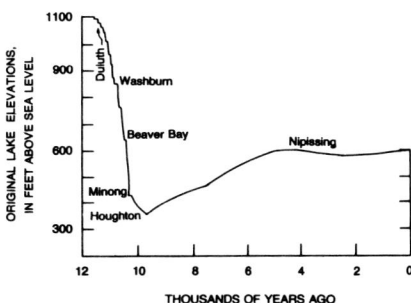

LAKE-LEVEL CHANGES in the Lake Superior basin. Declining lake levels are chiefly due to progressive uncovering of lower outlets as the ice retreated northward. Rising lake levels are due to uplift of outlets by crustal rebound. (Fig. 57)

are well displayed on Isle Royale above the present shoreline, actually existed at an elevation more than 150 feet lower than the present level of lake Superior.

Lake Duluth was the first and highest

glacial lake to fill a major part of the Lake Superior basin during retreat of the last ice (fig. 58). The ice sheet was in a period of rather rapid retreat when Lake Duluth formed, but it then must have slowed down because the immediately following lakes did not extend much farther northeast than Lake Duluth did, although lake levels fell nearly 500 feet between the levels of Lake Duluth and Lake Beaver Bay.

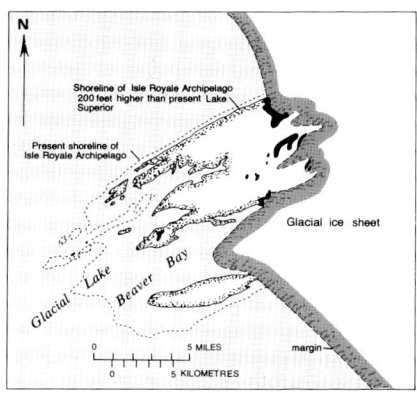

GLACIAL ICE MARGIN during deposition of recessional moraines on Isle Royale and prior to complete northeastward retreat of ice from the island. (Fig. 59)

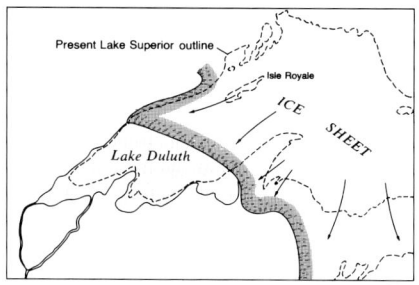

LAKE DULUTH and its contemporary ice border, about 11,000 years ago. (Fig. 58)

The ice front forming the north margin of the earlier lakes probably remained south of Isle Royale until about the time of Lake Beaver Bay or a little later, when it retreated to a position straddling Isle Royale west of Lake Desor (fig. 59). Abundant glacial till was deposited upon the newly emergent west end of the island, and the ice front remained stable long enough to build across the island the previously described complex of recessional moraines. Shorelines formed by the glacial lake associated with this ice front are found on the west end of the island where they now occur about 200 feet above the level of Lake Superior.

Subsequent renewed and complete retreat of the ice margin from Isle Royale was rapid enough that only a minor amount of glacial till was deposited on the central and eastern parts of the island. When the ice margin reached the north edge of the Lake Superior basin, Lake Minong was formed, and the entire basin was filled for the first time since the readvance of the ice sheet (fig. 60). Lake Minong marked a relatively stable period in the history of the basin, and its beaches are among the best developed on Isle Royale. In fact, it was on Isle Royale that evidence for the existence of Lake Minong was first recognized.

An even lower lake, Lake Houghton, followed Lake Minong. Then, during a long period, slowly rising lake levels controlled largely by crustal uplift re-

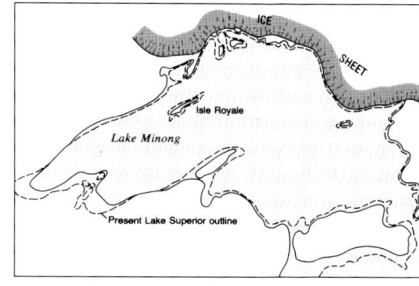

LAKE MINONG and contemporary ice border along the north shore of the Lake Superior basin, about 10,500 years ago. (Fig. 60)

sulted about 5,000 years ago in the Nipissing lake stage at 605 feet above sea level (fig. 57), and finally the present Lake Superior evolved at 602 feet (fig. 61).

On Isle Royale, beaches of the postglacial lakes ancestral to Lake Superior are best developed on the southwest end of the island, where abundant glacial deposits provided easily worked materials for beach construction (fig. 62). Most abandoned beaches are not readily accessible to the hiker, but the trail from Siskiwit Bay to the Island mine crosses rather prominent Nipissing beaches about one-quarter mile from Siskiwit Bay. The trail from the Siskiwit Bay campground to Feldtmann Ridge follows a Nipissing beach terrace through a clearing marking the site of an abandoned Civilian Conservation Corps camp and then climbs to a Minong beach, which it follows for about 2½ miles.

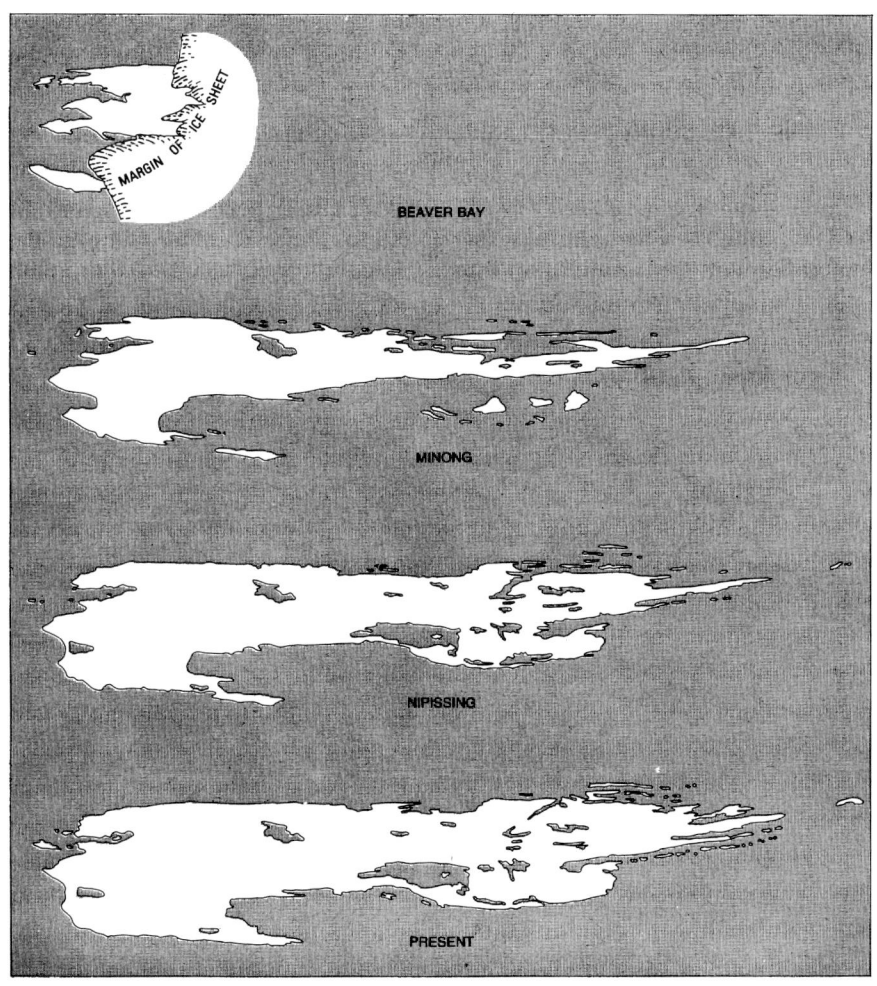

BIRTH AND GROWTH OF ISLE ROYALE during selected postglacial lake stages. (Fig. 61)

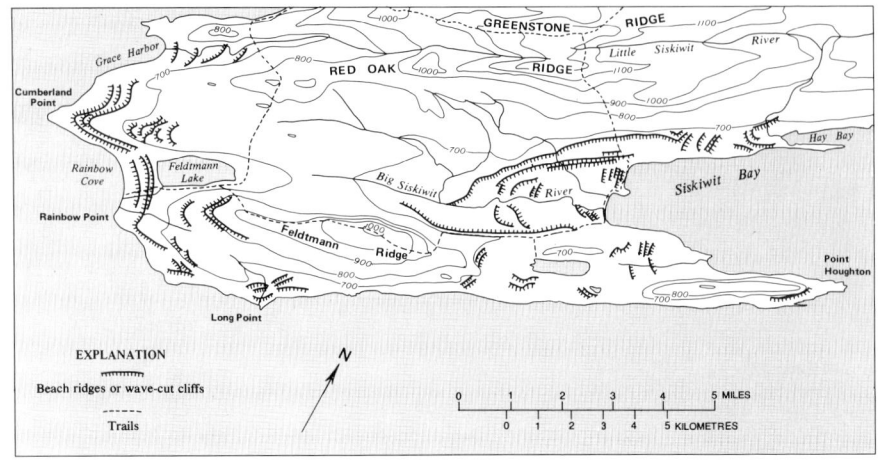

PROMINENT ABANDONED SHORELINE FEATURES on the southwest end of Isle Royale. (Fig. 62)

Among the best developed Minong and Nipissing beaches are those between Rainbow Cove and Feldtmann Lake; these are accessible from the trail between the cove and the lake (figs. 63, 64). Indeed, these beaches dam

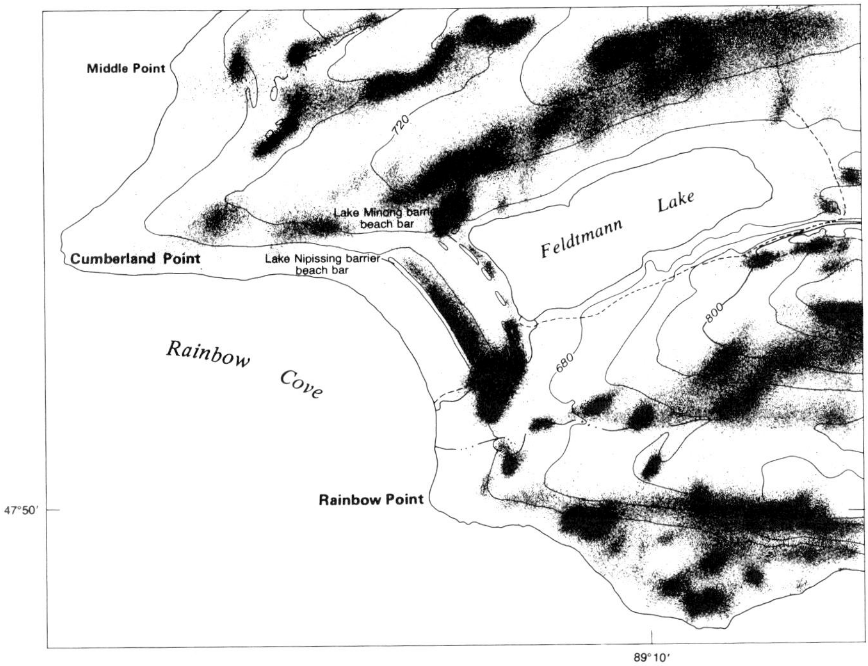

LAKE MINONG AND LAKE NIPISSING BARRIER BEACH BARS between Rainbow Cove and Feldtmann Lake. (Fig. 63)

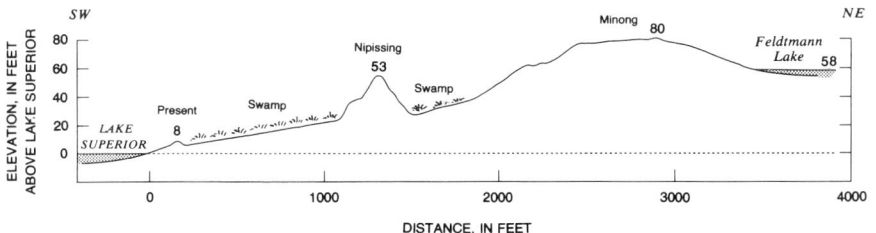

LAKE MINONG AND LAKE NIPISSING BARRIER BEACH BARS (profile) between Rainbow Cove and Feldtmann Lake. (Fig. 64)

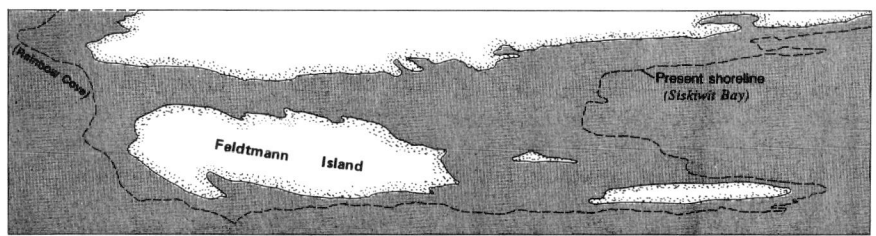

Shoreline of Isle Royale just prior to Lake Minong stage.

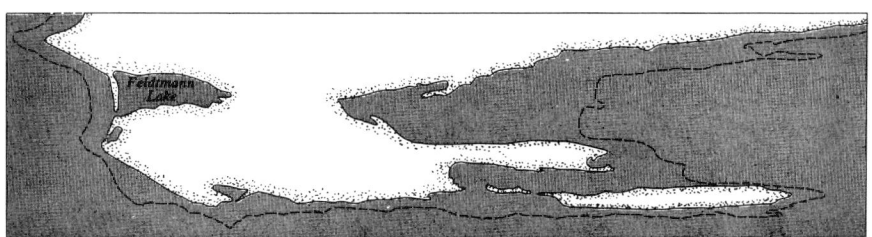

Formation of barrier bar to close off Feldtmann Lake during Lake Minong stage.

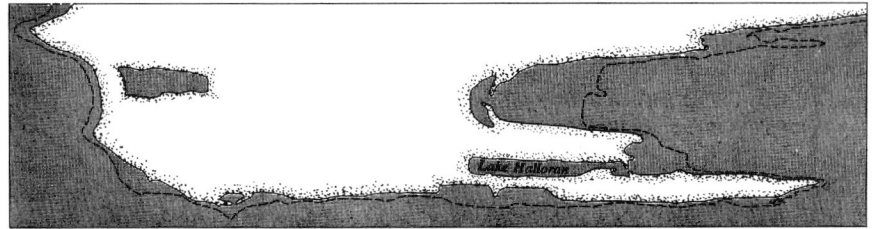

Formation of barrier bar to close off Lake Halloran during Lake Nipissing stage.

FELDTMANN AND HALLORAN LAKES development. (Fig. 65)

Feldtmann Lake. Just prior to the time of Lake Minong, the lowland between Rainbow Cove and Siskiwit Bay was covered by a glacial lake at a higher elevation, and Feldtmann Ridge was a separate island (fig. 65). As the water level fell to that of Lake Minong, the central part of this lowland emerged from the lake; however, Siskiwit Bay still extended inland about 5 miles more than it does today, and a similar bay on Lake Minong occupied the site of Feldtmann Lake. The Rainbow Cove area was exposed to the full force of storm waves from the southwest, and Lake Minong was at a stable elevation long enough for those waves to construct a barrier beach bar across the mouth of "Feldtmann Bay" and isolate it as a lake. Other beach bars were subsequently built at the lower levels of the Nipissing stage, including one that formed Lake Halloran.

On the northeast end of Isle Royale, where glacially transported debris is limited and abandoned beaches are less evident, wave-cut features in the bedrock mark ancient shorelines (fig. 66).

MONUMENT ROCK, a stack associated with the shoreline of postglacial Lake Minong. Adjacent wave-cut cliff is just out of photograph to right. National Park Service photograph by Tom Haas. (Fig. 67)

WAVE-CUT SHORELINE FEATURES including cliffs, stacks, and an arch. (Fig. 66)

Prominent examples are Monument Rock, a stack associated with the Minong shoreline north of Tobin Harbor (fig. 67), and an arch cut through a narrow ridge crest on Amygdaloid Island (fig. 68), probably associated with the shoreline of the Nipissing stage. Suzy's Cave, on the north side of Rock Harbor about 2 miles west of Rock Harbor Lodge, may also be associated with the Nipissing shoreline.

Except for wave-cut shoreline features, changes in the topography of Isle Royale have been very slight since the ice left. Materials derived from slope erosion are transported only very short distances, owing to low stream gradients and innumerable beaver dams. Bogs and swamps occupy much of the alluvial valley bottoms.

WAVE-CUT ARCH on Amygdaloid Island, associated with the shoreline of postglacial Lake Nipissing. (Fig. 68)

SOME MINERALS OF SPECIAL INTEREST

Many interesting minerals occur on Isle Royale in addition to the basic rock-forming minerals that make up the bulk of the lava flows and other rocks on the island. Among these are the native copper that played such an important part in the early explorations of the island and chlorastrolite, the official state gem of Michigan.

These minerals as a group are known as secondary minerals, ones introduced into the rocks after the rocks themselves were formed. They chiefly fill holes in the volcanic rocks and the conglomerates or form veins filling fractures. Some of these minerals, especially those occurring as amygdules in the volcanic rocks, are quite attractive when polished and long have been sought by collectors, although collection has been restricted since Isle Royale became a national park. Fortunately for the collector, these minerals, except for chlorastrolite, are equally or more abundant on the Keweenaw Peninsula or elsewhere in the Lake Superior region.

Considerable evidence obtained during mining activities on the Keweenaw Peninsula indicate that the copper and most other secondary minerals were deposited from solutions that percolated upward through the rocks. We can only speculate upon the ultimate source of the copper and other elements in the mineralizing solutions, but one of the more generally accepted theories is that those elements were "sweated" out of the lower part of the pile of volcanic rocks after those rocks were warped downward in the Lake Superior syncline, into a region of higher temperature and pressure. The elements then migrated upward and were deposited as minerals in open spaces higher in the rock sequence. If the source of the mineralizing solutions was indeed the deeper part of the volcanic pile, then the source also would have been closer to the Keweenaw Peninsula, for the axis, or deepest part of the Lake Superior syncline, is closer to the Keweenaw Peninsula than to Isle Royale. This asymmetry of the syncline may have been one of the more important factors in producing economically valuable deposits of copper on the peninsula but not on Isle Royale.

Among the many secondary minerals on Isle Royale, those occurring most often are barite, calcite, chlorite, copper, datolite, epidote, laumontite, natrolite, prehnite, chlorastrolite (pumpellyite), quartz (including agate), and thomsonite. Only a few of special interest on Isle Royale are described further in this report. Information on the others can be obtained from many sources, including the booklet "Rocks and Minerals of Michigan" (Poindexter and others, 1965).

NATIVE COPPER—WIDESPREAD BUT NOT ABUNDANT

Copper, like most other metals, most commonly occurs in nature bound up with other elements into minerals such as chalcocite(Cu_2S), chalcopyrite ($CuFeS_2$), and cuprite (Cu_2O), and such minerals form the bulk of copper ores. Of all the metals that do occur in the native or metallic state, copper is by far the most common; however, major concentrations of native copper are rare, and so it is noteworthy that the most important deposits known in the world are those of the Keweenaw Peninsula.

Economically valuable concentrations or deposits of native copper on the peninsula can be divided into two broad groups—lode deposits and fissure deposits. The lode deposits comprise mineralized conglomerate beds and the vesicular tops of lava flows; in each case the primary porosity

COPPER NUGGET weighing 5,720 pounds found at a depth of 16½ feet in a pit dug by prehistoric Indians at the site of the Minong mine. Note the uneven surface resulting from attempts to remove sections for implements (Burton collection, Detroit Public Library). (Fig. 69)

of the rock or presence and continuity of open spaces was a factor in determining where the ore would be deposited. The fissure deposits are along fracture zones that generally cut across the beds. Some of the fissure deposits contained great masses of metallic copper as much as hundreds of tons in weight. Extensive Indian mining pits on the fissures later led prospectors to most of the known deposits of this type. However, such deposits, rich as they were, have been much less important than the lower grade but vastly larger lode deposits, which have produced about 98 percent of the total copper mined in the Native Copper district, about 5,400,000 tons.

Native copper is widely distributed on Isle Royale, but mineralization was apparently too weak to develop large lode deposits. Except for the Island mine, which was on a conglomerate lode, most of the prospects and short-lived mines on the island were opened on small fissure deposits. In the fissure occurrences, such as at the Minong, Siskiwit, and most other mines and prospects, native copper occurs in nodules and irregular masses in highly altered rock in fracture zones a few inches to many feet wide. Several large masses more than a ton in weight were found at the Minong mine (fig. 69); these were rare and most pieces mined were probably similar to that in figure 70, or smaller. Although approximately 250 tons of copper were produced from the Minong mine, the copper was too sparse and widely distributed to be mined profitably.

At the Island mine, near the west end of Siskiwit Bay, native copper occurs in the matrix of a conglomerate, the only known occurrence of this type on Isle Royale. The specimen illustrated (fig. 71), however, is much richer in copper than average for the mine, and this mine too was a financial failure.

NATIVE COPPER MASS from the Minong mine. This is perhaps typical of specimens from the fissure deposits on Isle Royale, but larger than average. (Fig. 70)

In other places in the park, most notably on the chain of islands south of Rock Harbor, veins of quartz, prehnite, and calcite contain scattered grains of copper. Native copper also occurs in amygdules in lava flows, especially in prehnite amygdules.

CONGLOMERATE WITH NATIVE COPPER from the Island mine. The specimen has been sawed and slightly polished so that the copper reflects light and shows up as light-colored irregular-shaped patches. Specimen is 8 centimetres wide. (Fig. 71)

Rakestraw (1966) has described the historic mining ventures on Isle Royale and what the visitor can see at some of the abandoned mine sites.

CHLORASTROLITE— MICHIGAN'S STATE GEM

Isle Royale has long been famous as the home of chlorastrolite, known informally in rock-collecting and lapidary circles as "Isle Royale greenstone." This usage of "greenstone" should not be confused with the use of the same term for a volcanic rock with a greenish hue, such as makes up Greenstone Ridge on the island. In 1972 the governor of Michigan signed a bill designating chlorastrolite as the official state gem.

Chlorastrolite, meaning "green star stone," occurs as amygdules or cavity fillings in certain of the lava flows on Isle Royale. When weathered out of the lava flows, it can be found on some of the island beaches as pea-sized pebbles, generally greenish in color. When polished, either by wave action on the beaches or artificially, the "greenstones" generally exhibit a distinctive and attractive mosaic or segmented pattern, sometimes referred to as "turtleback" (fig. 72). The polished stones also commonly are chatoyant—the property of having a luster resembling the changing luster of the eye of a cat. Chatoyancy is probably best known in the gemstone called tiger eye and is a property of translucent material that contains fibrous structures capable of scattering light. The grouping together of bundles of such fibers produces the mosaic pattern of the "greenstones."

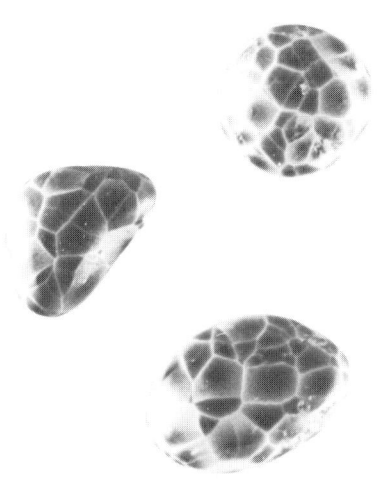

CHLORASTROLITE AMYGDULES showing characteristic segmented pattern. (Fig. 72)

Chlorastrolite was first discovered on Isle Royale and named and described by C. T. Jackson and J. D. Whitney in 1847. Long afterward it was found to be the same material as another mineral, pumpellyite, first described from the Keweenaw Peninsula in 1925 and named for Raphael Pumpelly, a pioneer student of the minerals of the Keweenawan copper deposits. The material from the peninsula was described in much greater detail than that from the island, and the name "pumpellyite" became deeply entrenched in the world mineralogical literature long before it was realized that the material from both areas was mineralogically the same. Consequently, pumpellyite has been adopted as the only valid name for the mineral species, although chlorastrolite is still useful as a term to designate the variety with the peculiar crystal habit of the Isle Royale "greenstone." Pumpellyite is common in many parts of the world, but the chlorastrolite variety is apparently rare outside of Isle Royale.

PREHNITE—THE LITTLE PINK PEBBLES

Prehnite is an abundant secondary mineral in some lava flows on Isle Royale and elsewhere in the Lake Superior region. It occurs as amygdule fillings, crosscutting veins, and as a replacement of earlier minerals or rock. Most of the prehnite has a characteristic pale-green to white color, but that in amygdules is commonly light to dark pink or variously mottled in pink and green. The pink prehnite superficially resembles thomsonite, with which it has commonly been confused; most of the so-called thomsonite from Isle Royale is actually pink prehnite, including material from Thomsonite Beach on the north side of the island. The prehnite does not develop the spectacular patterns and color variations present in gem-quality thomsonite, which explains why "Isle Royale thomsonite" has always been considered to be of inferior quality. Nevertheless, the prehnite amygdules polish nicely and are attractive in themselves.

The pink prehnite amygdules, which most commonly range in size from ½ to 1 centimeter, are more resistant to erosion than the volcanic rock matrix within which they have formed. As a result, beach pebbles containing them often have a knobby appearance, the amygdules projecting above the general surface of the matrix (fig. 73). Where the prehnite amygdules weather completely free from the matrix, they may make up a fair percentage of the fine gravel on beaches near the prehnite-bearing outcrops.

The amygdules in figure 74 are typi-

PREHNITE AMYGDULES in beach pebble derived from lava flow. Long dimension of pebble is 5 centimetres. (Fig. 73)

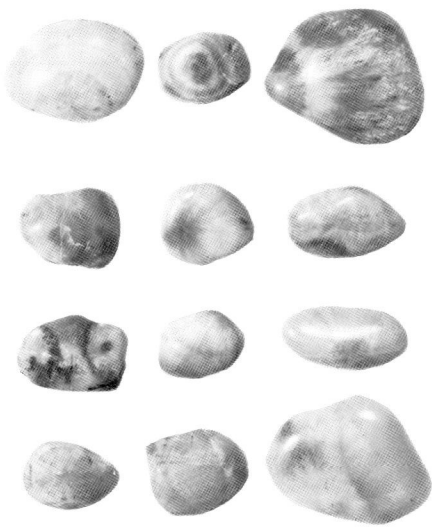

PREHNITE AMYGDULES—note radiating fibrous structure with development of "eyes." (Fig. 74)

cal and illustrate the radiating fibrous habit, with the occasional development of "eyes," that has probably been a factor in their confusion with thomsonite. The pink color of the prehnite is due to internal reflections from finely disseminated native copper inclusions, and the color intensity is related to the distribution, abundance, and grain size of the inclusions.

AGATE—AN ARRAY OF COLORS

Perhaps the best known gemstones of the Lake Superior region are agates. Although they are most plentiful near where they weather from the enclosing rocks, glacial transport has caused them to be widely distributed, and there are few pebble beaches where one cannot be found.

Agate is a subvariety of chalcedony (fibrous quartz) with a distinct banding in which successive layers differ in color and in degree of translucency. It most commonly originates as a cavity filling in volcanic rocks either as amygdules in volcanic flows or as irregular-shaped masses in volcanic tuffs; both types occur in the Lake Superior region, including Isle Royale.

One of the most striking examples of an agate-bearing lava flow in the park is the Amygdaloid Island Flow. This flow, exposed only on Amygdaloid Island, has rather abundant rounded or almond-shaped agates with a characteristic flesh-pink color and commonly massive quartz or vuggy centers (fig. 75). Agates from an individual flow tend to be somewhat similar in appearance; for example, those from the Amygdaloid Island Flow are pinkish, whereas those from the Long Island Flow tend to be bluish. In general, agates from the rocks on Isle Royale are

AGATE AMYGDULES in fine-grained trap from Amygdaloid Island Flow. (Fig. 75)

paler and have less color contrast between color bands than those found in some other parts of the Lake Superior region.

The volcanic tuff that overlies the Greenstone Flow also contains numerous agates with a pink or red cast (fig. 76). Unlike the ovoid agates typically occurring in the lava flows, these agates tend to have very irregular shapes similar to the so-called thunder-egg agates occurring in welded tuffs of the Columbia River Plateau in the northwestern United States, and they probably had a similar origin. The welded tuffs formed from volcanic ash that fell in a hot plastic condition that permitted it to be fused or welded into a generally cohesive mass. The agates themselves are interpreted as chalcedony deposited in cavities formed by pockets of gases accumulating during cooling, shrinkage, and partial crystallization of the tuff.

Because of their hardness, agates are more resistant to erosion than the enclosing rock and tend to become concentrated in gravel and other surficial

AGATE typical of that in the volcanic tuff above the Greenstone Flow. (Fig. 76)

deposits both ancient and modern. Thus agates are present in the conglomerates interbedded with the lava flows on the island and in the Copper Harbor Conglomerate, as well as in the modern surficial deposits. Most of the agates on beaches near conglomerate outcrops have, therefore, gone through two cycles of weathering and release from their host rock—first from the volcanic rock in which they originated and then from the conglomerate in which they were subsequently incorporated. In addition to beach agates derived from nearby bedrock, some have been winnowed from glacial till by wave action and may have been transported a long distance from their original source; they commonly are quite different in some way from the majority of the agates on a given beach.

WHAT THE FUTURE HOLDS?

As we pick up an agate from the beach and admire its color and pattern, we can reflect upon the enormous amount of change that has taken place since it originated as an amygdule in a lava flow a billion or so years ago. The time from the volcanic eruptions to the coming of the glacier that may have brought that agate to its resting place, however, was itself enormous, even in the geologic sense—nearly one-fourth the age of the earth itself. And although things are relatively quiet on Isle Royale at the present time, geologically speaking, changes will continue to take place slowly but certainly. Storm waves will continue to batter the shoreline cliffs, wearing them back. Winter frost will loosen rock fragments on the hillsides for the summer rains to wash downslope. The streams will deposit silt in the landlocked lakes until they are filled and cease to exist as bodies of water.

These continuing changes take place so slowly that they are nearly imperceptible to us. They are quite different from the cataclysmic eruption of a volcano or the steady advance and envelopment of an ice sheet. But perhaps this is just a quiet moment in the continuing geologic history of Isle Royale. While renewed volcanic activity is unlikely, many scientists believe that we are now in an interglacial epoch and that massive ice sheets will again form on the Canadian Shield to move inevitably southward, again modifying the earth's surface as they grind their way over it. Perhaps the future glacier might even bring some new agates to the beach for another visitor to admire.

A WORD OF THANKS

Alfred C. Lane opened his geologic report on Isle Royale with the following words: "The progress of knowledge is like the growth of a coral reef; each generation builds upon that which has been left behind by those who have gone before." So it is with this report. I have drawn upon so many sources of information in presenting this geologic story of Isle Royale that it is impossible to acknowledge the individual contributions of each. This is especially true of the historical material and of numerous works that have delineated the regional geologic framework within which I have placed Isle Royale. For those who wish to delve further into various aspects of the history and geology of this part of the Lake Superior region, several pertinent references, briefly annotated, are listed in the bibliography; these in turn will lead to additional source materials.

As part of the current work, a new multicolor geologic map of Isle Royale National Park has also been prepared (Huber, 1973c). It has been published separately and is at a scale of 1:62,500 (approximately 1 in.=1 mi). At that scale it naturally shows much more detail than any of the maps in this report and illustrates more graphically many of the geologic features described herein. It is thus highly recommended as a supplement to this report. Several reports more technical than this one also have been prepared, one of them with Roger G. Wolff (Huber, 1971, 1973a, b; Wolff and Huber, 1973).

The study leading to the preparation of this report and the companion geologic map and other reports was conceived and carried out as a cooperative project between the Geological Survey and the National Park Service with two goals in mind: to take advantage of the geologic research potential of the park and to help interested visitors understand its geologic story. During the four summers that I spent mapping the geology of Isle Royale, the hospitality and enthusiastic support of every member of the Park Service staff contributed immeasurably to the success of the project. In the geologic mapping I was ably assisted by Robert J. Larson in 1966, Harrison T. Southworth in 1967, Charles E. Bartberger in 1968, and David R. Chivington in 1969.

Henry R. Cornwall and Walter S. White aided considerably through numerous discussions that drew upon their broad knowledge of Keweenawan geology. Most of the text and many illustrations for the discussion of the postglacial lake stages are based largely on the work of William R. Farrand (figs. 56, 57, 58, 60) and the unpublished Ph.D. dissertation (Univ. of Michigan, 1932) of George M. Stanley (figs. 55, 61, 64).

DEFINITION OF TERMS

An attempt has been made to minimize the use of technical terms in this volume, but some jargon is inevitable in any discussion of technical matters. Most of the strictly geologic terms that are apt to be stumbling blocks are defined where they first appear in the text; for ease of reference, some of them have been summarized in this short glossary. Geologic time terms have not been included. (See the section on "Time as a Geologic Concept.")

Alluvial fan. A sloping fan-shaped mass of loose rock material deposited by a stream where it emerges from an upland into a broad valley or plain.

Alluvium. A general term for clay, silt, sand, and gravel deposited by running water, such as a stream.

Amygdule. A gas cavity or vesicle in volcanic rock that has been filled with secondary minerals such as calcite or quartz.

Basalt. The most common type of volcanic rock, generally fine grained, dark, heavy, and with a silica (SiO_2) content of about 50 percent or less.

Basement. The assemblage of metamorphic and igneous rocks underlying the stratified rocks in any particular region.

Breccia. A rock consisting of consolidated rock fragments like a conglomerate, except that most of the fragments are angular rather than rounded.

Chlorastrolite. A variety of the mineral pumpellyite, a complex hydrous calcium silicate. (See section on "Chlorastrolite—Michigan's State Gem.")

Chlorite. A green complex silicate mineral, similar in structure to the micas and usually of secondary origin.

Cleavage. The tendency of a mineral to break along definite planes, producing smooth surfaces.

Columnar jointing. Parallel prismatic columns, polygonal in cross section, formed by contraction during cooling in some lava flows.

Conglomerate. A rock, the consolidated equivalent of gravel.

Drumlin. A streamlined hill of glacial till elongate in the direction of flow of the glacier that formed it. The crag-and-tail variety is streamed out behind a bedrock knob.

Erratic (glacial). A rock fragment, usually of a large size, that has been transported from a distant source, especially by the action of glacial ice.

Fault. A break in rock strata along which displacement has occurred.

Felsite. A fine-grained light-colored volcanic rock, generally reddish. (See section on "Volcanic Rocks" for fuller description.)

Flood basalt (plateau basalt). A basaltic lava flow or successive lava flows that spread over very large areas from fissure eruptions.

Formation. In geology, an assemblage of rocks that have some character in common that distinguishes them from adjacent rocks.

Greenstone. (1) Informal name for the mineral chlorastrolite (pumpellyite); (2) a volcanic rock with a greenish hue, such as makes up Greenstone Ridge.

Igneous rock. A rock formed by solidification of hot molten material, either at depth in the earth's crust (plutonic) or erupted at the earth's surface (volcanic).

Keweenawan Supergroup. A sequence of geologic formations of which the Portage Lake Volcanics and the Copper Harbor Conglomerate are a part; Keweenawan is also used informally to denote that part of Precambrian time during which the rock sequence was deposited.

Metamorphic rock. Rock changed materially in composition or appearance by heat, pressure, or infiltrations at depth in the earth's crust.

Moraine. An accumulation of glacial till with an initial topographic expression of its own, usually a ridge, unrelated to the surface upon which it lies. Several varieties have been recognized; for example, a recessional moraine is one formed at the margin of a receding ice sheet during a pause in its recession.

Ophite. A volcanic rock with a mottled texture, often exhibiting a knobby surface when weathered. (See section on "Volcanic Rocks" for fuller description.)

Pegmatite. A volcanic rock in which elongate plagioclase laths produce a matted appearance. (See section on "Volcanic Rocks" for fuller description.)

Playa. A level or nearly level area that occupies the lowest part of a completely closed basin and that is covered with water at irregular intervals, forming a temporary lake.

Porphyrite. A volcanic rock with a texture produced by well-defined plagioclase crystals scattered through a finer grained matrix. (See section on "Volcanic Rocks" for fuller description.)

Prehnite. A calcium aluminum silicate mineral. (See section on "Prehnite—The Little Pink Pebbles.")

Pyroclastic rock. Rock formed by consolidation of ash or other fragmental material explosively ejected from a volcano.

Rhyolite. A light-colored commonly reddish volcanic rock with a high silica (SiO_2) content of 70 percent or more.

Sedimentary rock. Rock formed by consolidation of sediment deposited at the surface of the earth, chiefly in, or through the action of, water.

Silica (SiO_2). Occurs as the natural mineral quartz, including various fine-grained varieties such as chert, chalcedony, agate, and others. It also occurs in most rock-forming minerals (silicates) and is a major factor in the chemical classification of volcanic rocks.

Stratigraphy. The arrangement of rock strata, especially as to geographic position and chronologic order of sequence.

Syncline. In simple form, a concave-upward, or troughlike, fold in stratified rock.

Talus. An accumulation of coarse angular rock fragments derived from and resting at the base of a cliff or very steep slope.

Till. Glacially transported material deposited directly by ice, without transportation or sorting by water.

Trap. A fine-grained dark-colored volcanic rock. (See section on "Volcanic Rocks" for fuller description.)

Tuff. Rock formed from consolidation of volcanic ash.

Vesicle. Small cavity formed by the expansion of bubbles of gas during solidification of a volcanic rock.

REFERENCES AND ADDITIONAL READING

Agassiz, Louis, 1850, Lake Superior—its physical character, vegetation, and animals, compared with those of other and similar regions; with a narrative of the tour, by J. Elliot Cabot: Boston, Gould, Kendall and Lincoln, 428 p. Reprinted 1970 (American Environmental Studies), New York, Arno Press.

For the history buffs, a marvelous narrative of Agassiz' expedition along the north shore of Lake Superior nearly as far west as the Pigeon River, with frequent geologic observations.

Bird, J. B., 1972, The natural landscapes of Canada; a study in regional earth science: Toronto, Wiley Pub. Canada Limited, 191 p.

A description of the physical landscape and its origin, including a very readable summary of the glaciation of the Canadian Shield.

Courter, E. W., 1972, The chlorastrolite; now Michigan's official gem: Conglomerate, v. 33, no. 2, p. 3, 8, 10, 12.

The nature and history of chlorastrolite, from its discovery to its new honor.

Dorr, J. A., Jr., and Eschman, D. F., 1970, Geology of Michigan: Ann Arbor, Univ. Michigan Press, 476 p.

Excellent overview of the geology of Michigan, for the general reader.

Farrand, W. R., 1969, The Quaternary history of Lake Superior: Internat. Assoc. Great Lakes Research, 12th Conf., Great Lakes Research, 1969, Proc., p. 181–197.
Summary of the glacial and postglacial lakes in the Lake Superior basin and their changing outlets.

Flint, R. F., 1971, Glacial and Quaternary geology: New York, John Wiley & Sons, 892 p.
Standard text with broad coverage of continental glaciation in North America.

Foster, J. W., and Whitney, J. D., 1850, Report on the geology and topography of a portion of the Lake Superior Land District in the State of Michigan; Pt. 1, Copper lands: U.S. 31st Cong., 1st sess. House Executive Doc. 69, 224 p.
Primarily of historical interest with sketches of early explorations in the Lake Superior region and the initiation of copper mining, including descriptions of early activities on Isle Royale.

Griffin, J. B., ed., 1961, Lake Superior copper and the Indians— miscellaneous studies of Great Lakes prehistory: Michigan Univ. Mus. Anthropology, Anthropol. Paper 17, 189 p., 33 pl.
Includes extensive data on prehistoric activities of Indian copper miners on Isle Royale.

Halls, H. C., 1966, A review of the Keweenawan geology of the Lake Superior region, *in* The earth beneath the continents—a volume of geophysical studies in honor of Merle A. Tuve: Am. Geophys. Union Geophys. Mon. 10 (Natl. Acad. Sci.–Natl. Research Council Pub. 1467), p. 3–27.
Summary of the correlation of individual groups of Keweenawan rocks around the Lake Superior basin.

Huber, N. K., 1971, Pink copper-bearing prehnite from Isle Royale National Park, Michigan: Earth Science, v. 24, no. 1, p. 9–14.
Data on pink prehnite and discussion of its common confusion with thomsonite.

——1973a, Glacial and postglacial geologic history of Isle Royale National Park, Michigan: U.S. Geol. Survey Prof. Paper 754–A, 15p.
Comprehensive discussion of material simplified for this report.

——1973b, The Portage Lake Volcanics (middle Keweenawan) on Isle Royale, Michigan: U.S. Geol. Survey Prof. Paper 754–C, 32p.
Detailed description of the volcanic rock sequence on Isle Royale.

——1973c, Geologic map of Isle Royale National Park, Keweenaw County, Michigan: U.S. Geol. Survey Misc. Geol. Inv. Map I–796, scale 1:62,500.
Details of geology not possible to show on smaller scale maps in this report.

Kelley, R. W., and Farrand, W. R., 1967, The glacial lakes around Michigan: Michigan Geol. Survey Bull. 4, 23 p.
Excellent summary for the general reader.

Lane, A. C., 1898, Geological report on Isle Royale, Michigan: Michigan Geol. Survey, v. 6, pt. 1, 281 p.
First comprehensive report on the geology of Isle Royale includes descriptions of exploration activities on the island through 1896.

Mech, L. D., 1966, The wolves of Isle Royale: National Park Service Fauna Series 7, 210 p.
A fascinating account with emphasis on the predator-prey relationship of the wolf and moose.

Poindexter, O. F., Martin, H. M., and Bergquist, S. G., 1965, Rocks and minerals of Michigan [5th ed.]: Michigan Geol. Survey Bull. 2,

101 p.
Introduction to the origin, location, and basic uses of rocks and minerals found in Michigan.

Rakestraw, Lawrence, 1965, Historic mining on Isle Royale: Isle Royale Nat. Hist. Assoc., 20 p.
Summary of post-Indian mining and exploration activities on Isle Royale, 1843-92.

U.S. Geological Survey, 1957, Topographic map of Isle Royale National Park (culture revised 1970): scale 1:62,500.
Available in contour or shaded-relief edition.

White, W. S., 1960, The Keweenawan lavas of Lake Superior, an example of flood basalts: Am. Jour. Sci. (Bradley volume), v. 258-A, p. 367-374.
The rationale behind the interpretation of the Portage Lake Volcanics as flood basalts.

—————1968, The native-copper deposits of northern Michigan, *in* Ridge, J. D., ed., Ore deposits of the United States, 1933-1967: New York, Am. Inst. Mining, Metall., and Petroleum Engineers, v. 1, p. 303-326.
One of the more recent summaries of the nature and origin of Michigan's native-copper deposits.

Wolff, R. G., and Huber N. K., 1973, The Copper Harbor Conglomerate (middle Keweenawan) on Isle Royale, Michigan, and its regional implications: U.S. Geol. Survey Prof. Paper 754-B, 15 p.
Detailed description of the Copper Harbor Conglomerate on Isle Royale including discussion of its depositional environment.

The real significance of wilderness is a cultural matter. It is far more than hunting, fishing, hiking, camping, or canoeing; it has to do with the human spirit.
Sigurd F. Olson, "The Spiritual Aspects of Wilderness," 1956

We at Avery Color Studios thank you for purchasing this book. We hope it has provided many hours of enjoyable reading.

Learn more about Michigan and the Great Lakes area through a broad range of titles that cover mining and logging days, early Indians and their legends, Great Lakes shipwrecks, Cully Gage's Northwoods Readers (full of laughter and occasional sadness), and full-color pictorials of days gone by and the natural beauty of this land. Beautiful note stationery is also available.

For a free catalog, please call 800-722-9925 in Michigan or 906/226-3338, or tear out this page and mail it to us. Please tape or staple the card and put a stamp on it.

PLEASE RETURN TO:

Avery
COLOR STUDIOS

P.O. Box 308
Marquette MI 49855

CALL TOLL FREE
1-800-722-9925

Your complete shipping address:

_____

_____
_____
_____

_____

Fold, Staple, Affix Stamp and Mail

Avery
COLOR STUDIOS

P.O. Box 308
Marquette MI 49855